T0295144

Einstein's Entanglement

Einstein's Entanglement

Bell Inequalities, Relativity, and the Qubit

W. M. STUCKEY
MICHAEL SILBERSTEIN
AND
TIMOTHY McDEVITT

OXFORD
UNIVERSITY PRESS

OXFORD
UNIVERSITY PRESS

Great Clarendon Street, Oxford, OX2 6DP,
United Kingdom

Oxford University Press is a department of the University of Oxford.
It furthers the University's objective of excellence in research, scholarship,
and education by publishing worldwide. Oxford is a registered trade mark of
Oxford University Press in the UK and in certain other countries

Published in the United States of America by Oxford University Press
198 Madison Avenue, New York, NY 10016, United States of America

British Library Cataloguing in Publication Data
Data available

Library of Congress Control Number: 2024933326

ISBN 9780198919674

DOI: 10.1093/9780198919698.001.0001

Printed and bound by
CPI Group (UK) Ltd, Croydon, CR0 4YY

Cover image: The qubit from Einstein's relativity dispels Einstein's spooks. ©T.D. Le

MIX
Paper | Supporting
responsible forestry
FSC® C013604

Preface

> The basic motivation is just to learn how Nature works. What's really going on? Einstein said it very nicely. He's not interested in this detailed question or that detailed question. He just wanted to know what were God's thoughts when He created the world.
>
> Anton Zeilinger (2019)

The Nobel Prize in Physics is typically awarded for discoveries about physical reality that answer Zeilinger's question [8], "What's really going on?" However, Alain Aspect, John Clauser, and Anton Zeilinger were awarded the 2022 Nobel Prize in Physics "for experiments with entangled photons, establishing the violation of Bell inequalities" that have created tremendous confusion among physicists and philosophers about "how Nature works." Igor Salom wrote [10]:

> The 2022 Physics Nobel Prize was not quite like any other. While the Nobel prizes in physics are always of interest to the physics community, by a rule, they are merely a matter of curiosity for the general public. However, the 2022 Nobel award should pertain to all of us, irrespective of the profession, and remind us that it's been a time to rethink our basic worldviews.

Had he lived to see these experimental results, Albert Einstein might well have questioned his belief that "God reveals himself in the lawful harmony of all that exists" [2]. In other words, Aspect, Clauser, and Zeilinger's discoveries have left us wondering, more so than ever, "What were God's thoughts in creating the world?" [Obviously, "God" or "Deity" as used herein has no necessary religious connotation.]

In fact, after Clauser did his experiment he was shocked by the result [9]:

> I kept asking myself, "What have I done wrong? What mistakes have I made in this?"

But, he found no mistakes and had to accept that the violation of Bell inequalities as predicted by quantum mechanics was correct. About this, Clauser said [9]:

I was, again, very saddened that I had not overthrown quantum mechanics because I had, and to this day still have, great difficulty in understanding it.

And Clauser is not alone, other Nobel Laureates in Physics have likewise proclaimed "nobody understands quantum mechanics." As the experimental results proving that Nature does indeed violate Bell inequalities in accord with quantum mechanics trickled in over the years, physicists and philosophers began debating this mystery of quantum entanglement.

Ironically, it was Einstein himself (with Boris Podolsky and Nathan Rosen) who introduced the mystery of quantum entanglement in 1935. In this book we will explain why quantum entanglement has been called "the greatest mystery in physics" [1] and why some believe it suggests "the ways of a Deity who is, if not evil, at least extremely mischievous" [7, p. 221]. Indeed, there is a widespread belief in foundations of quantum mechanics that quantum entanglement renders the two pillars of modern physics, quantum mechanics and special relativity, fundamentally incompatible. Worse, many believe quantum entanglement forces us to accept one or more of the following:

- Quantum mechanics is incomplete.
- The world works according to "spooky actions at a distance."
- Causes from the future create effects in the present.
- There is "superdeterministic" causal control of experimental procedures.
- People can correctly disagree on the outcome of one and the same experiment.
- A single experimental measurement can produce all possible outcomes.

In order to show you how all of these conclusions can be avoided, we must ask two things of you.

First, if you are a colleague in foundations of quantum mechanics, we ask that you read the entire presentation even though the book is written at the level of a general reader who is only familiar with basic concepts from an introductory course in physics, e.g., frequency of outcomes, probability, vectors, velocity, force, energy, electromagnetism, etc. Since we are not assuming the reader has taken a course in quantum mechanics or special relativity, our presentation of these topics may strike our colleagues as primitive. However, the way we need the reader (general and expert alike) to think about quantum entanglement and special relativity requires a very specific approach to these topics that experts may not have seen since their undergraduate studies.

Second, we need readers (general and expert alike) to set aside, in Einstein's language, their *constructive bias*, i.e., the belief that any solution to the mystery of quantum entanglement must "allow us to *trace* the *causal mechanisms* that *explain mechanistically* the occurrence of" the Bell-inequality-violating quantum correlations [5]. The key to escaping all of the unacceptable consequences of Nature's violation of Bell inequalities articulated above resides in abandoning the belief that explanation must be in terms of dynamical laws and/or mechanistic causal processes (causal mechanisms).

If you can set aside your constructive bias and read the book completely, the payoff is a possible solution to the greatest mystery in physics that does not violate locality, statistical independence, intersubjective agreement, or unique experimental outcomes. Additionally, our proposed solution reveals that quantum mechanics is complete and totally compatible with special relativity. In what proves to be yet another irony associated with this mystery, we will solve Einstein's mystery of quantum entanglement the same way Einstein solved the mystery of length contraction with his theory of special relativity in 1905.

The mystery of length contraction at the end of the 19th century had to do with the fact that everyone measured the same value for the speed of light c, regardless of their relative motions (light postulate). Physicists assumed light was a wave in a medium called the luminiferous aether and they were able to account for the light postulate if objects shrank along the direction of their motion through the aether (length contraction). So, the constructive explanation they sought was:

$$\text{aether} \rightarrow \text{length contraction} \rightarrow \text{light postulate.}$$

Accordingly, a causal mechanism per the aether leading to length contraction was the greatest mystery in physics of that time.

Even Einstein sought a constructive explanation of the light postulate before giving up, saying [4]:

By and by I despaired of the possibility of discovering the true laws by means of constructive efforts based on known facts. The longer and the more despairingly I tried, the more I came to the conviction that only the discovery of a universal formal principle could lead us to assured results.

The "universal formal principle" that Einstein used in lieu of "constructive efforts" was the *relativity principle*:

The laws of physics (including their constants of Nature) are the same in all inertial (non-accelerated) reference frames.

Essentially, his principle solution to the mystery of length contraction flipped the explanatory sequence. That is, he first used the relativity principle to justify the light postulate because c was a constant of Nature according to Maxwell's equations of electromagnetism. That means measurements of c must produce the same result in all inertial reference frames, which includes those in uniform relative motion. He then showed how the light postulate gives rise to length contraction, so his principle solution to the mystery of length contraction was:

relativity principle → light postulate → length contraction.

You can see how he simply bypassed causal mechanisms altogether. As we will explain in this book, the exact same opportunity presented itself to Einstein regarding the mystery of quantum entanglement. In a nutshell, here is what he could have done along these same lines to prevent himself from ever positing a mystery of quantum entanglement to begin with.

Einstein's 1921 Nobel-Prize-winning discovery, the photoelectric effect equation, employed a constant of Nature called Planck's constant h. So, Einstein could have pointed to the Stern–Gerlach experiment in 1922 as experimental evidence for the fact that everyone measures the same value for Planck's constant h, regardless of their relative spatial orientations. Let us call this the *Planck postulate* in analogy with the light postulate. Since inertial reference frames are related by relative spatial orientation as well as uniform relative motion, the relativity principle justifies the Planck postulate just like it justifies the light postulate.

As we will establish, quantum information theorists have shown that the Hilbert space formalism of quantum mechanics, to include its entangled states, follows from the information-theoretic counterpart to the Planck postulate called *Information Invariance & Continuity*. Putting all of this together, we see that if Einstein had produced the reconstruction of quantum mechanics from the Planck postulate à la quantum information theorists, he would have had the following principle solution to the mystery of quantum entanglement before it was even a mystery:

relativity principle → Planck postulate → quantum entanglement.

As a principle solution, this does not violate locality, statistical independence, intersubjective agreement, or the uniqueness of experimental outcomes. But, we should give Einstein the benefit of the doubt for his adherence to constructive explanation in quantum mechanics. After all, he died nine years before John Bell published his inequality, and seventeen years before the first experimental violation of it by Clauser.

If Einstein had been aware of the physics behind the 2022 Nobel Prize in Physics, he may well have abandoned his "constructive efforts" regarding quantum mechanics and turned to his beloved relativity principle yet again. We will never know that, but you can decide for yourself if Einstein's approach to Einstein's entanglement makes sense after reading this book.

Notes for the Reader

For those readers who know quantum mechanics, what little quantum formalism we do use in this book is almost exclusively the Hilbert space vector formalism. Therefore, you will not see any pictures of 'electron clouds' like the atomic or molecular orbitals in physical chemistry. Instead, the quantum state (wavefunction) is a vector in Hilbert space whose basis vectors represent the possible measurement outcomes for the measurement in question.

Just as physical vectors (such as angular momentum $\vec{L}$) can be projected on the unit vectors $\hat{x}$, $\hat{y}$, or $\hat{z}$ to find their components (L_x, L_y, L_z) along the x, y, or z axes in space, respectively, the quantum state $|\psi\rangle$ can be projected on each unit measurement outcome vector (e.g., the measurement outcome vectors $|+\rangle$ and $|-\rangle$ for the measurement outcomes +1 and –1, respectively) to find its components (ψ_+ and ψ_-) along the possible measurement outcomes in Hilbert space. A component (probability amplitude) is then squared ($|\psi_+|^2$ and $|\psi_-|^2$) to find the probability that the quantum state will produce that specific measurement outcome (+1 or –1). That is the extent to which we use quantum formalism in this book.

Indeed, we will only be concerned with the two-outcome Hilbert space like the example we just gave, i.e., the quantum bit or *qubit*. Quantum information theorists have reconstructed quantum mechanics showing how all the higher-dimensional Hilbert spaces are built from the two-dimensional qubit Hilbert space. So, in order to solve the mystery of quantum entanglement we just need to understand why the qubit is 'weird'. What we will show in this book is that the 'weirdness' of the qubit (called *superposition*) is the result of the Planck postulate (Information Invariance & Continuity) as justified by the relativity principle (explained above).

In addition to a very simple mathematical presentation, we have also simplified the philosophical presentation that one often encounters in scholarly papers on quantum entanglement. Where we have used philosophical terminology we have provided an intuitive definition. While the book is written at a level suitable for the general reader, we have supplied references to scholarly work for our physics and philosophy colleagues in foundations of quantum mechanics.

If you are only interested in the physics and want to avoid the philosophy as much as possible, you may skip Chapters 1, 3, 5, and 9 (except 9C, where the qubit for the double-slit experiment is introduced). On the other hand, if you are only interested in the philosophy and want to avoid detailed physics, you may skip Chapters 6, 7, and 8. Finally, if you are only interested in a conceptual presentation and consequences, you may skip Chapters 3, 5, 6, 7, and 8. Of course, the general or expert reader interested in learning as much as possible about our thesis and its consequences should read the entire book.

We also point out that we will make extensive use of direct quotes throughout the book. We feel that it is best to provide the exact words of those we cite, so readers may make their own inferences and/or more easily search for the broader context of the idea. At times the quotes are rather lengthy, but we believe this is preferable to even longer paraphrased text.

The goal of this book is to argue for a principle alternative to the constructive explanations of quantum mechanics. And, to borrow the words of Sean Carroll [8]:

Truly understanding quantum mechanics will only happen when we put ourselves on the entanglement side, and we stop privileging the world that we see and start thinking about the world as it actually is.

We will show you how "the world actually is," i.e., we will answer "What's really going on?" according to Einstein's relativity principle. That includes solving the mystery of quantum entanglement without violating locality, statistical independence, intersubjective agreement, or unique experimental outcomes, and without altering quantum mechanics. According to Einstein, "The whole of science is nothing more than a refinement of everyday thinking" [3], but as you will learn in this book, to understand how the world actually is requires more than a mere "refinement of everyday thinking." The world we will reveal in this book, i.e., the world 'God created' impartially per the relativity principle, will have profound consequences for your everyday thinking. What we will reveal accords nicely with the recent findings of Tomonori Matsushita and Holger Hofmann [6]:

Our results show that the physical reality of an object cannot be separated from the context of all its interactions with the environment, past, present, and future, providing strong evidence against the widespread belief that our world can be reduced to a mere configuration of material building blocks.

In short, be prepared to say goodbye to reality as you currently understand it.

References

[1] A. D. Aczel, *Entanglement: The Greatest Mystery in Physics*, Basic Books, New York, 2002.

[2] J. Baggott, *What Einstein meant by "God does not play dice"*, 2018. https://www.britannica.com/story/what-einstein-meant-by-god-does-not-play-dice.

[3] A. Einstein, *Physics and Reality*, Journal of the Franklin Institute, 221 (1936), pp. 349–382.

[4] A. Einstein, *Autobiographical notes*, in Albert Einstein: Philosopher-Scientist, P. Schilpp, ed., Open Court, La Salle, IL, 1949, pp. 3–94.

[5] D. Maltrana, M. Herrera, and F. Benitez, *Einstein's theory of theories and mechanicism*, International Studies in the Philosophy of Science, 35 (2022), pp. 153–170.

[6] T. Matsushita and H. F. Hofmann, *Dependence of measurement outcomes on the dynamics of quantum coherent interactions between the system and the meter*, Physical Review Research, 5 (2023), p. 033064.

[7] T. Maudlin, *Quantum Non-Locality and Relativity*, Wiley-Blackwell, Oxford, 2011.

[8] NOVA Season 46 Episode 2, *Einstein's Quantum Riddle*, 2019. https://www.pbs.org/video/einsteins-quantum-riddle-ykvwhm/.

[9] NOVA The Fabric of The Cosmos, *The Illusion of Distance and Free Particles: Quantum Entanglement*, 2013. https://www.youtube.com/watch?v=ZNedBrG9E90.

[10] I. Salom, *2022 Nobel Prize in Physics and the End of Mechanistic Materialism*, Phlogiston, 31 (2023). In Press. https://arxiv.org/abs/2308.12297.

Acknowledgments

We would like to thank the anonymous reviewers at Oxford University Press for their very thoughtful and helpful comments. For a close reading of large portions of the text with helpful comments we owe a special debt of thanks to our colleagues Emily Adlam, Jeff Bub, Časlav Brukner, Lucien Hardy, Peter Lewis, David Mermin, Markus Müller, John Ranck, Perry Rice, Jesse Waters, and Ken Wharton. We also thank our students Justin Cedeno, Justin Cosgrove, Isaac Kraenbring, Ryan Smith, and Kaitlynn Yorgey for reading select material and providing feedback. For the cover art and creating the app "Mermin's Challenge" we thank Tuyen Le. And finally, we would like to thank our editor at OUP, Sonke Adlung, for support and guidance.

Contents

0

Albert's Mysterious Quantum Gloves Experiment

I would not call [quantum entanglement] *one* but rather *the* characteristic trait of quantum mechanics, the one that enforces its entire
departure from classical lines of thought.

Erwin Schrödinger (1935)

In this pre-introductory chapter, we present the mystery of quantum entanglement and our principle solution thereto in the form of a fictitious experiment,
a parable of sorts. We will refer to this fictitious experiment all the way through
Chapter 4, before talking about the mystery of quantum entanglement exclusively in terms of actual quantum mechanics. Since we are writing the book
to be accessible to the general reader, we have provided this opening parable
as a conceptual introduction to the more detailed material that follows in later
chapters. Essentially, we want to ease the reader (general and expert alike) into
our way of thinking about the mystery, as well as the reason it has remained
unsolved for nearly 90 years. Our goal overall in this chapter is to put you into
the proper frame of mind to appreciate our principle solution to the greatest
mystery in physics à la Einstein.

0.1 Entanglement and Conservation

According to Schrödinger's quote [2], and as we will explain shortly in our
parable, quantum entanglement is what makes quantum mechanics quantum instead of classical. However, entanglement in general simply means that
information obtained about one member of a pair of objects allows you to
know something about the other member of that pair, i.e., the information
you have about them is *correlated* [3]. Obviously, this is not mysterious per se.

For example, suppose Albert sends one glove of a pair to Alice in London
and the other to Bob in Hong Kong. If Alice opens her box and finds it is the

Einstein's Entanglement. W. M. Stuckey, Michael Silberstein, and Timothy McDevitt, Oxford University Press.
© Oxford University Press (2024). DOI: 10.1093/9780198919698.003.0001

right-hand glove, and she knows Albert is sending gloves from the same pair, then she knows immediately that Bob will find or has found that his box contains the left-hand glove. Obtaining information about the handedness of her glove in conjunction with information about the pair immediately provided information about the handedness of Bob's glove even though they were very far apart. While it is unconventional, we might say handedness is conserved as follows.

Numerically, we could give right a value of +1 and left a value of −1 so that right and left add to zero. Thus, Alice and Bob would say a pair of gloves has total handedness of zero and conservation of handedness dictates that when Alice obtains a measurement outcome of right (+1), Bob's handedness measurement outcome must be left (−1) in accord with conservation of handedness. That is, Alice and Bob's handedness outcomes are *anti-correlated* due to conservation of handedness. Again, this is not typically how anyone would explain such an obvious result, but our motive for thinking this way will be clear when we consider *quantum* gloves in what follows. The bottom line is, there is no mystery concerning the entangled outcomes for ordinary (classical) gloves.

To create a mystery, we need to add another property to the gloves, e.g., color, and allow for a means of combining the two properties called *superposition*. True, there is no way to measure 20% handedness and 80% color or 30% handedness and 70% color, etc., for classical gloves, but this will make sense for quantum gloves as we will see. In the following parable, we start by adding the property of color, then we introduce the superposition of properties leading to the mystery of quantum entanglement. We will also share our solution of that mystery exactly as Einstein might have solved it, if he had known what we know today about quantum entanglement.

0.2 Albert's Classical Gloves Experiment

Suppose Albert challenges his two students Alice and Bob to collect data and explain their results in the following experiment. Let the classical (ordinary) gloves of our previous example come in two colors (black and white) and suppose that Alice and Bob can only measure one property of the gloves in each trial of the experiment. That is, in each trial of the experiment Alice can choose to measure either handedness or color of her glove and Bob can choose to measure either handedness or color of his glove. Why are we imposing this restriction?

While we can certainly measure both color and handedness of a classical glove with a mere glance, we cannot simply 'look at' a quantum particle to obtain information about it, we have to perform an actual quantum measurement. And as it turns out, there is a limit to how much simultaneous information one can possess about a quantum particle. In the fictitious quantum gloves experiment that follows, quantum color and quantum handedness cannot be known simultaneously (they are called *complementary variables*). In order that the classical gloves experiment mirrors the quantum gloves experiment, Alice is only allowed to measure one property, color or handedness, in each trial of the classical gloves experiment. The same goes for Bob.

So, Albert provides Alice and Bob each a glove measurement device such that a glove enters Alice's (Bob's) measurement device and Alice (Bob) can press either the color button or the handedness button to obtain the corresponding measurement outcome (Figure 0.1). Notice that their measurement options form a discrete (individually separate) set. Put a pin in that.

After conducting many trials of the experiment, Alice and Bob convene to analyze their data. What they discover first is that in any trial of the experiment when they both happened to measure color, they always obtained the same outcome; specifically, half the time both were white and half the time both were black. And, when they both happened to measure handedness, they always obtained opposite outcomes, i.e., half the time Alice's outcome was right and Bob's was left and half the time Alice's outcome was left and Bob's was right. Alice and Bob have thus identified that the gloves are entangled with respect to *each* of the properties of color and handedness.

So, if they were to continue this experiment in the future and Alice were to measure the color of her glove and obtain white, she would know

Fig. 0.1 A source sends a pair of entangled non-quantum (classical) gloves to Alice and Bob's measurement devices. Each measurement device has two measurement settings (two buttons) and two outcomes for each (on top) for Alice or Bob's measurements of a classical glove. In each trial of the experiment, Alice may choose to measure color or handedness (not both) for her glove, ditto for Bob.

immediately that if Bob were to also measure the color of his glove in that trial, he would obtain white. If Alice were to measure the handedness of her glove and obtain left, she would know immediately that if Bob were to also measure the handedness of his glove in that trial, he would obtain right. And so on.

As with conservation of handedness, Alice and Bob could also say color is conserved. For example, they could assign white a value of +1 and black a value of –1 and say that a pair of gloves has total color of +2 or –2, i.e., no pair of gloves has total color of zero (one glove is white and the other is black). That is, Alice and Bob's color outcomes are correlated due to conservation of color. Consequently, we can say that classical gloves are entangled with respect to color and handedness because of conservation of color and handedness.

Given these conservation principles for classical gloves, Alice and Bob explain their entangled (correlated or anti-correlated) outcomes by assuming Albert is sending them gloves from the same pair of white or black gloves in each trial of the experiment. Additionally, they reason that half the time Albert sends the gloves from a black pair and half the time he sends the gloves from a white pair, and half the time he sends Alice (Bob) the right-hand glove and half the time he sends Alice (Bob) the left-hand glove. All of the results described so far can be understood according to that simple explanation. They just need to test their hypothesis against their data for the outcomes of different measurements.

That is, if this is in fact what Albert is doing, then they should find absolutely no correlation between color and handedness in those trials when they happen to choose to measure different properties. Checking their data for all the trials when Alice measured color and obtained white and Bob measured handedness, they do in fact find that Bob obtained right 50% of the time and left 50% of the time. And, for all the trials when Alice measured color and obtained black and Bob measured handedness, they do in fact find that Bob obtained right 50% of the time and left 50% of the time. They also confirm that the outcomes for Alice's measurements of handedness with Bob's measurements of color are equally uncorrelated.

To summarize their findings, the same properties are totally correlated (color) or totally anti-correlated (handedness) due to conservation of color and handedness, while different properties are totally uncorrelated. After performing their rigorous data analysis, they excitedly take their explanation to Albert who responds, "You got it, that's what I am doing!" He also commends them for thinking like physicists and framing their findings in terms of conservation principles.

0.3 Albert's Quantum Gloves Experiment

In this example with classical gloves, the properties of color and handedness are distinct. If Bob asked Alice, "What color is your glove?" and Alice answered "Left," Bob would be very confused. Right and left are only outcomes of a handedness measurement while black and white are only outcomes of a color measurement. Therefore, the two measurements of color and handedness cannot be combined. What would that even mean? But, suppose that it made sense to talk about Alice or Bob measuring 20% color and 80% handedness of their glove or 50% color and 50% handedness of their glove, etc. That is called *superposition*, and quantum mechanics says you can do it for quantum particles.

So, if Albert is sending quantum gloves, Alice and Bob have lots more options for their measurements. Since any such combination is possible, we are talking about a continuous set of measurement choices and our quantum gloves need to have the properties of quantum color and quantum handedness with corresponding quantum outcomes of Black or White and Left or Right, respectively, such that it makes sense to measure any combination of them.

In that case, Albert must give Alice and Bob each a new measurement device with a dial to make measurements over their continuous set of options, i.e., quantum color or quantum handedness or any superposition between (Figure 0.2). Notice that this is very different from making measurements on non-quantum (classical) gloves where the measurement options formed a discrete set and the measurement device had two buttons (Figure 0.1). The next concern is outcomes.

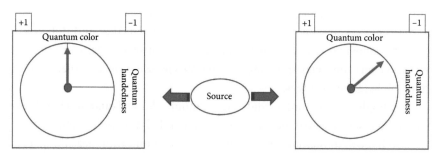

Fig. 0.2 A source sends a pair of entangled quantum gloves to Alice and Bob's measurement devices. Each measurement device has continuous measurement settings (dial direction in space) with two outcomes (±1 on top) for Alice or Bob's measurements of a quantum glove. In this trial of the experiment, Alice (on the left) is measuring quantum color while Bob (on the right) is measuring a superposition of quantum color and quantum handedness.

How will we express the outcome of a superposition measurement, e.g., 30% quantum color and 70% quantum handedness? WhiRig? BlaLef? In quantum experiments, we have uniform binary outcomes for all such measurements, e.g., up or down, vertical or horizontal, +1 or −1, etc. Let's use ±1, so that quantum color and quantum handedness measurement outcomes might be interpreted to mean the same number for the same quantum color (+1 is White and −1 is Black) and the same number for different quantum handedness (+1 is Right for Alice and Left for Bob, −1 is Left for Alice and Right for Bob). The bottom line is that each measurement device only outputs +1 or −1 in each trial of the experiment, regardless of its measurement setting (dial orientation). Having given Alice and Bob each a quantum glove measurement device, Albert again challenges them to collect data and explain their results in the quantum version of the previous experiment.

As with the classical gloves experiment, in those trials when Alice and Bob both happen to measure quantum color, they both get +1 half the time and they both get −1 half the time. And, in those trials when they both happen to measure quantum handedness, they both get +1 half the time and they both get −1 half the time. In fact, this correlation holds true for *any* trial of the experiment when they both happen to set their measurement devices to the same setting (same dial direction), quantum color or quantum handedness *or any superposition thereof.*

As with the non-quantum (classical) gloves, we can explain these entangled outcomes using a conservation principle. What might we call the quantity that is being conserved? It was easy for the classical gloves because color and handedness don't mix, so we simply had two distinct conservation principles, i.e., one for color and one for handedness. But now our conserved quantity is quantum color or quantum handedness or any combination thereof, so we need a generic name for what is being conserved when Alice and Bob both make the same measurement (same dial direction in space). Alice decides to call this generic property *spin*, since solving this puzzle is making their heads spin. Bob laughs and agrees.

Accordingly, we now understand that quantum color and quantum handedness are just spin measurements in two different dial directions. When Alice and Bob happen to measure spin in the same direction (both measure quantum color or both measure quantum handedness or both measure spin in another common direction), then they always obtain the same ±1 outcome because of the conservation of spin. That is, the source is simply emitting the quantum gloves with the same spin in any direction in any given trial of the experiment in accord with the conservation of spin. This is totally analogous to the classical measurements of color and handedness for the classical gloves. So, just like in the classical case, Alice and Bob are ready to test their conservation of spin explanation in trials where they decided to choose dial settings in different directions, i.e., in trials where they make different spin measurements. What might Alice and Bob expect to find in those trials?

Recall that for classical gloves the properties of color and handedness are distinct and Albert was always sending gloves from the same pair, so Alice and Bob found no correlation between color and handedness outcomes when they made different measurements. But, this is not true for quantum gloves where quantum color and quantum handedness are just spin measurements in two particular dial directions and spin is conserved in all directions. So, Alice and Bob do expect to find some correlation between their outcomes when they make different spin measurements, and that leads Alice to reason as follows.

Alice considers all the trials when she measured quantum color and got +1 and Bob measured at a dial setting making a particular angle θ with respect to quantum color, e.g., $\theta = 60°$ (Figure 0.3). She knows that Bob would have obtained +1 if he had measured quantum color because of the conservation of spin, so his quantum glove must have an underlying (hidden) value of +1 in the quantum color direction. Accordingly, when Bob measures spin at an

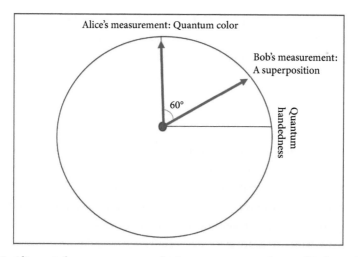

Fig. 0.3 Alice sets her measurement device to quantum color and Bob sets his measurement device to a superposition of quantum color and quantum handedness that makes an angle $\theta = 60°$ with respect to Alice's setting. The settings are shown here on the same dial, but they actually exist on each of Alice's and Bob's measurement devices.

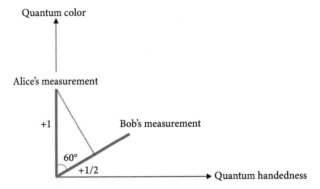

Fig. 0.4 Alice says that when she measures quantum color and obtains +1 and Bob measures spin at 60° relative to quantum color, Bob's outcome should be $\frac{1}{2}$ because of conservation of spin.

angle θ with respect to quantum color, he should simply find the projection of +1 along his new measurement direction, i.e., +1 times $\cos(\theta) = \cos(60°) = \frac{1}{2}$ (Figure 0.4).

Unfortunately, Bob tells Alice that he only ever obtained ±1 in that dial direction, just like any other dial direction, just like Alice. Alice complains that Bob is probably making sloppy measurements. Annoyed, Alice tells Bob to average his results in that direction, since it is possible for a collection of +1 and −1 outcomes to *average* to $\cos(\theta)$ for some θ. Bob checks his outcomes at that angle in those trials where Alice obtained +1 and sure enough, for every −1 outcome he did have three +1 outcomes, so his outcomes did average to (Table 0.1)

$$\frac{-1+1+1+1}{4} = \frac{1}{2} = \cos(60°).$$

Alice says Bob should be embarrassed getting only +1 or −1 in those trials when he is supposed to be getting $\frac{1}{2}$, even if those results do average to $\frac{1}{2}$. But Bob is very confident in his experimental technique, so he believes his measurement results are accurate. Feeling insulted, Bob challenges Alice to look at her results in those trials corresponding to *his* +1 outcomes. After all, he reasons, if Alice had measured in his direction he knows she would have obtained +1 for her outcome, so *she* should be the one obtaining $(+1)\cos(\theta) = \cos(60°) = \frac{1}{2}$ (Figure 0.5). And sure enough, Alice checks her data and finds that her outcomes are only +1 or −1 in those trials and they average to ... $\frac{1}{2}$ (Table 0.1). Alice and Bob have no explanation for this symmetrical 'average-only' conservation, so they ask Albert for some hints.

Table 0.1 Example collection of eight data pairs when Alice and Bob's quantum glove measurement settings differ by 60° (the two columns on the left). Alice partitions the data according to her ±1 results to show that Bob's measurement outcomes only *average* the required $\pm\frac{1}{2}$ for conservation of spin (the middle two columns). But, when Bob partitions the data according to his ±1 results he can show it is *Alice's* measurement outcomes that only average the required $\pm\frac{1}{2}$ for conservation of spin (the two columns on the right).

Data		Alice's partition		Bob's partition	
Bob	Alice	Bob	Alice	Bob	Alice
+1	+1	+1	+1	+1	+1
−1	+1	−1	+1	+1	−1
−1	−1	+1	+1	+1	+1
+1	−1	+1	+1	+1	+1
−1	−1	−1	−1	−1	+1
+1	+1	+1	−1	−1	−1
−1	−1	−1	−1	−1	−1
+1	+1	−1	−1	−1	−1

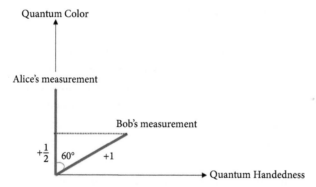

Fig. 0.5 Bob says when he measures spin and obtains +1 and Alice measures quantum color at 60° relative to his spin measurement, Alice's outcome should be $\frac{1}{2}$ because of conservation of spin.

0.4 Albert's Three Rules and John's Inequality

Albert admits that he himself has no explanation for the behavior of the entangled quantum gloves, even though he has been thinking about it since he and his friends Boris and Nathan introduced the mystery of quantum entanglement in 1935. While many solutions have been proposed for this mystery, Albert says he doesn't like any of them. He wants an explanation that obeys three very specific rules, and while all explanations to date conform to the first rule, they all violate the second and/or third rule.

The first rule says there has to be some behind-the-scenes mechanistic causal process or dynamical law (hidden variables) to account for the response of the measurement device to the quantum glove. He calls these hidden variables a *constructive* account of the measurement and all attempts to explain his experiment obey this rule. The second and third rules deal with restrictions on how the causal mechanism of the hidden variables works, and that is where violations of the rules occur.

Specifically, the second rule says the hidden variables must not require information sent faster than light from Bob's (Alice's) measurement settings to Alice's (Bob's) measurement device to determine Alice's (Bob's) measurement outcome. He calls this *locality* and says these first two rules mean that "physics should represent a reality in time and space, free from spooky actions at a distance" [1].

Finally, the third rule says the hidden variables do not function according to the measurement setting until the quantum glove arrives at the measurement device because the information about the measurement setting is not available to the quantum glove until it enters the measurement device. Relatedly, Alice and Bob's measurement choices must be random with respect to any of the hidden variables, i.e., it cannot be the case that some hidden variables are measured more/less often than others. He calls this *statistical independence* and says this rule is so obvious that many physicists do not even bother to point it out. Alice and Bob agree that Albert's rules are reasonable, indeed they would not even consider an explanation that did not obey his three rules.

On their way out of Albert's office, a man who overheard their discussion politely draws Alice and Bob aside, saying, "It sounds like Albert has given you the task of explaining his quantum gloves experiment according to his very strict rules." Alice and Bob nod in agreement and ask the man if he knows the answer. The man tells them his name is John and he was here to show Albert a proof concerning Albert's quantum gloves experiment.

As it turns out, John has a relationship (in the form of an inequality) that must be satisfied in any experiment that obeys Albert's three rules, and John can show that Albert's quantum gloves experiment violates the relationship. For example, suppose Alice picks any two of her measurements $[A_1, A_2]$ and Bob picks any two of his measurements $[B_1, B_2]$. Given that the outcome of any measurement is either +1 or −1, in those trials when Alice measured A_1

and Bob measured B_1 the average of the A_1 outcome (a_1) times the B_1 outcome (b_1), $\langle a_1 b_1 \rangle$, cannot be larger than +1 or smaller than –1. The same goes for any of the other three combinations $\langle a_1 b_2 \rangle$, $\langle a_2 b_1 \rangle$, or $\langle a_2 b_2 \rangle$. Albert's three rules then dictate that the quantity $\langle a_1 b_1 \rangle + \langle a_1 b_2 \rangle + \langle a_2 b_1 \rangle - \langle a_2 b_2 \rangle$ cannot be larger than +2 or smaller than –2. John calls this the CHSH quantity after his friends Clauser, Horne, Shimony, and Holt.

Alice and Bob discuss John's reasoning and quickly conclude that it is sound. John then says, "Go through all your data and select $[A_1, A_2, B_1, B_2]$ such that the angle between A_1 and B_1 is 45°, the angle between A_1 and B_2 is 45°, the angle between A_2 and B_1 is 45°, and the angle between A_2 and B_2 is 135° (Figure 0.6). After you've done that, compute the CHSH quantity for the outcomes of those measurements and see what you get." John smiles knowingly and says, "Excuse me, I have to show this to Albert now. It was nice meeting you." Alice and Bob thank John for his help and head off to undertake his task.

After painstakingly following John's instructions they are amazed at their result: their data give a value of $2\sqrt{2}$ for the CHSH quantity in violation of John's inequality. The conclusion is unavoidable: the constructive explanation per the first rule that Albert seeks for his quantum gloves experiment is impossible because the second and/or third rules that Albert demands must be violated in the quantum gloves experiment. Before heading to their respective dorms, Alice and Bob agree to carefully consider which of Albert's rules, two or three, they are willing to break and meet up to discuss the situation at lunch the next day. All in all, there doesn't seem to be much hope of satisfying Albert.

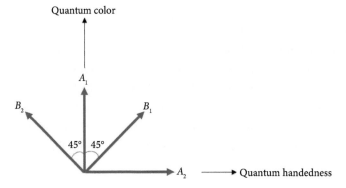

Fig. 0.6 A set of measurements that violate John's inequality.

0.5 Alice and Bob Solve the Mystery

At lunch the next day, Bob and Alice agree that, based on the history of such efforts, violating rules two or three will not produce a consensus solution to the mystery of quantum entanglement. Bob tells Alice, "John's inequality and Albert's three rules pertain to constructive explanations, so we need a different explanation altogether, one that will produce a consensus solution and that Albert will accept. But, what?" After a brief pause they have a simultaneous insight, pointing at each other they exclaim in unison, "We did this before!"

What they both recalled was an experiment Albert had them do some years ago that produced an analogous paradox. In that experiment, Alice and Bob were moving very fast relative to each other and had to measure the length of each other's meter sticks. Paradoxically, Alice found Bob's meter stick was shorter than hers, while Bob found Alice's meter stick was shorter than his (length contraction).

In the quantum gloves experiment, Alice says conservation of spin can only be satisfied on average in those trials where Bob makes a different measurement than she makes because his outcomes are always ± 1. That means Bob's outcomes can only *average* the required fraction ($\pm \cos(\theta)$) needed for exact conservation of spin according to her ± 1 data. The paradox arises because Bob can use *his* ± 1 data to make the exact same argument about Alice's ± 1 data. Maybe, they reason, they can solve the mystery of quantum entanglement like Albert solved the mystery of length contraction!

In that case, Alice and Bob failed to come up with a causal mechanism for length contraction, so they asked Albert for a hint. To their surprise, Albert told them he also tried and failed to develop a constructive explanation for length contraction, so he gave up in despair and developed a principle account. According to his principle explanation, length contraction follows from the fact that everyone must measure the same value for the speed of light c, regardless of their relative motions (an empirical fact called the *light postulate*). And the light postulate follows from the (very compelling) relativity principle, i.e., the laws of physics (including their constants of Nature) are the same in all inertial (non-accelerated) reference frames. That is, Albert's universally accepted principle solution to the mystery of length contraction was:

relativity principle → light postulate → length contraction.

Accordingly, Alice and Bob each end up with data that shows the other's meter sticks are shorter than their own. "So," Bob says, "maybe 'average-only' conservation can be explained the same way. But, what's the compelling principle?"

After a long pause, Alice erupts excitedly, "It's Albert's own relativity principle!" Bob asks how that can be true when they weren't moving at high speeds relative to each other, indeed they weren't moving relative to each other at all during the quantum gloves experiment. Alice points out that uniform relative motion is just one way to relate inertial reference frames. "Remember,;' she says, "we both measured ±1 for all dial directions in space, and inertial reference frames are also related by spatial rotations. So, it's just a matter of figuring out which universal constant of Nature is being found to have the same value in all those different inertial reference frames."

Bob realizes immediately that the constant they need must be Planck's constant h because it is *the* universal constant of Nature that distinguishes quantum mechanics from classical mechanics. That is, equations in quantum mechanics become equations in classical mechanics when $h = 0$. And, h also establishes a limit on the amount of simultaneous information one may obtain for a quantum particle. In information-theoretic terms, when $h = 0$ one obtains classical physics where there is no limit to the simultaneous information for particles.

This is analogous to the fact that relativistic equations become Newtonian equations when $c = \infty$. And, c also establishes a limit on the speed at which information can be exchanged. Therefore, the invariance of the value of h for measurements in different inertial reference frames, regardless of their relative spatial orientations, must be responsible for the ±1 spin outcomes in inertial reference frames related by spatial rotations. They decide to call this the *Planck postulate* in analogy with the light postulate, both being empirical facts justified by the relativity principle. Alice and Bob are now ready to take the following principle explanation of the quantum gloves experiment to Albert:

relativity principle → Planck postulate → 'average-only' conservation.

How could Albert refuse his very own relativity principle as an explanation of his mysterious quantum gloves experiment?

Just as they are ready to set off for Albert's office, Bob pauses and asks Alice, "Why do you think Albert didn't already propose this principle resolution to his paradox himself?" Alice reminds Bob that Albert much prefers constructive explanation and only proposed his principle explanation of length contraction after becoming convinced that he would never find an acceptable constructive explanation. Now that John has explained to Albert that his desired constructive explanation of this experiment is impossible, maybe Albert would turn again to a principle explanation. "But," Bob interrupts, smiling, "we beat him to it."

References

[1] M. Born, A. Einstein, and I. Born, *The Born Einstein Letters: Correspondence between Albert Einstein and Max and Hedwig Born from 1916 to 1955 with Commentaries by Max Born*, Macmillan Press, London, 1971. Translated by Irene Born.

[2] E. Schrödinger, *Discussion of Probability Relations between Separated Systems*, Mathematical Proceedings of the Cambridge Philosophical Society, 31 (1935), pp. 555–563.

[3] F. Wilczek, *Entanglement Made Simple*, Quanta Magazine, 28 April (2016).

1

Introduction: Nobody Understands Quantum Mechanics

> The necessity of the quantum in the construction of existence: out of what deeper requirement does it arise? Behind it all is surely an idea so simple, so beautiful, so compelling that when—in a decade, a century, or a millennium—we grasp it, we will all say to each other, how could it have been otherwise? How could we have been so stupid for so long?
>
> John Wheeler (1986)

The problem Wheeler alludes to in this 1986 passage [44] persists today, i.e., we do not understand quantum mechanics. Clauser's quote in the Preface stating that he wished he had overthrown quantum mechanics because "I had, and to this day still have, great difficulty in understanding it" [33] is just one of many. Murray Gell-Mann also won the Nobel Prize in Physics for his work in quantum physics and said [46, p. 144]:

> We all know how to use [quantum mechanics] and how to apply it to problems; and so we have learned to live with the fact that nobody can understand it.

Richard Feynman is yet another Nobel Prize winner in Physics for his work in quantum physics who said [21]:

> I think I can safely say that nobody understands quantum mechanics.

David Mermin is renowned for his work in quantum physics and he wrote [29]:

> Everybody who has learned quantum mechanics agrees how to *use* it. "Shut up and calculate!" There is no ambiguity, no confusion, and spectacular success. What we lack is any consensus about what one is actually *talking about* as one uses quantum mechanics. There is an unprecedented gap between the

Einstein's Entanglement. W. M. Stuckey, Michael Silberstein, and Timothy McDevitt, Oxford University Press.
© Oxford University Press (2024). DOI: 10.1093/9780198919698.003.0002

abstract terms in which the theory is couched and the phenomena the theory enables us so well to account for. We do not understand the *meaning* of this strange conceptual apparatus that each of us uses so effectively to deal with our world.

No consensus solution to this problem has been found despite the concerted efforts of many brilliant minds over many decades. It even plagued Einstein, despite the fact that he won his Nobel Prize in Physics for his work in quantum physics.

In fact, the problem of understanding quantum mechanics bothered Einstein so much that he questioned the validity of quantum mechanics altogether. Here is an excerpt from his letter to Max Born in 1947 [7]:

> I cannot seriously believe in [quantum mechanics] because the theory cannot be reconciled with the idea that physics should represent a reality in time and space, free from spooky actions at a distance.

Einstein's phrase "spooky actions at a distance" is a reference to quantum entanglement. [From now on entanglement will mean quantum entanglement unless specified otherwise.] Einstein relied on entanglement to argue that quantum mechanics is incomplete in a 1935 paper with Boris Podolsky and Nathan Rosen [16] (although Schrödringer didn't coin the term *entanglement* until after Einstein's paper). The Einstein, Podolsky, and Rosen (EPR) argument was simple and we will state it in terms of our quantum gloves experiment in Chapter 0.

Recall that when Alice measures quantum color, she can use that information to know immediately Bob's outcome for a quantum color measurement because the quantum gloves are entangled with respect to quantum color. And, when Alice measures quantum handedness, she can use that information to know immediately Bob's outcome for a quantum handedness measurement because the quantum gloves are also entangled with respect to quantum handedness. Therefore, it is possible to know with 100% certainty (probability = 1) the value of either of these properties without Bob's actually measuring them.

Assuming that Alice's measurement settings and outcomes are not instantly being conveyed to Bob's quantum glove so as to influence his measurement outcome (that is, assuming locality), EPR said quantum color and quantum handedness are "elements of reality" (realism). Since quantum mechanics does not allow for the simultaneous knowledge of quantum color and quantum handedness (they are complementary variables in quantum mechanics) with probability = 1, quantum mechanics is incomplete. Indeed, Einstein added this remark in his letter to Born [7]:

> I am quite convinced that someone will eventually come up with a theory whose objects, connected by laws, are not probabilities but considered facts [probability = 1], as used to be taken for granted until quite recently.

Lee Smolin went so far as to attribute our difficulty understanding quantum mechanics to its apparent incompleteness, writing [37, p. xvii]:

> I hope to convince you that the conceptual problems and raging disagreements that have bedeviled quantum mechanics since its inception are unsolved and unsolvable, for the simple reason that the theory is wrong. It is highly successful, but incomplete.

1.1 The EPR and EPR–Bell Paradoxes

Einstein died in 1955 probably still believing quantum mechanics needed to be completed by some "hidden variables" corresponding to the missing "elements of reality" responsible for the correlations of entangled particles. At the time, no one knew how to put Einstein's belief to the test, but that changed in 1964 with John Bell's paper [5] "On the Einstein Podolsky Rosen paradox." The paper opens:

> The paradox of Einstein, Podolsky and Rosen ... was advanced as an argument that quantum mechanics could not be a complete theory but should be supplemented by additional variables. These additional variables were to restore to the theory causality and locality.

Notice that Bell mentions Albert's first rule of constructive explanation ("causality") and his second rule of locality, but not the third rule of statistical independence. That is because violating statistical independence (or intersubjective agreement or the uniqueness of experimental outcomes, all discussed later) is not even an option for most physicists. However, as we will see, given that the mystery of entanglement remains unsolved after nearly 90 years, these other options are now being considered out of 'desperation' by some scholars in foundations of quantum mechanics.

In his paper, Bell showed that hidden ("additional") variables of the sort EPR believed were missing in quantum mechanics lead to empirical consequences at odds with quantum mechanics (Chapters 2 and 3). In terms of the quantum gloves experiment, Bell explored the consequences of those trials in which Alice and Bob made different measurements of the complementary variables and their superpositions, as explained in Chapter 0. Bell found that if certain reasonable assumptions are true (à la Albert's three rules in Chapter 0), the experiment would necessarily produce outcomes that satisfy an inequality like that in Chapter 0, i.e., the data must always produce a CHSH quantity between 2 and –2. The problem is, there are certain measurements of entangled particles that quantum mechanics predicts will violate Bell's inequality, e.g., they produce a CHSH quantity as large as $2\sqrt{2}$ or as small as $-2\sqrt{2}$.

Many experiments have since been conducted along the lines proposed by Bell, and all results to date vindicate quantum mechanics, i.e., the outcomes violate Bell's inequality exactly as predicted by quantum mechanics. The unavoidable conclusion is that any constructive solution (rule one) to the mystery of entanglement will necessarily violate locality and/or statistical independence (rules two and/or three). Consequently, Bell's work has been called "the most profound discovery of science" [45].

Concerning Bell's discovery, Tim Maudlin writes [28]:

> The experimental verification of violations of Bell's inequality for randomly set measurements at space-like separation is the most astonishing result in the history of physics. Theoretical physics has yet to come to terms with what these results mean for our fundamental account of the world.

To understand what the term "space-like separation" means and why it is important, consider a spacetime diagram. The vertical coordinate of a spacetime diagram is typically time, the horizontal coordinate is typically space, and an object's location plotted in space at various times creates what physicists call a *worldline* for the object (Figure 1.1). In the spacetime diagram for special relativity (Minkowski spacetime), the units for the time and space axes are usually chosen such that the worldline for a light pulse is angled at 45° (Figure 1.2). The relationships between locations in spacetime (called *events*) are categorized according to this so-called *lightcone structure*.

In Figure 1.2, Event C is inside the future lightcone of Event A and is said to be *timelike related* to Event A. Since Event C is inside the lightcone for Event A, a signal sent from Event A to Event C could be traveling slower than

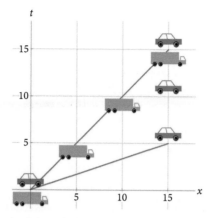

Fig. 1.1 A spacetime diagram of two vehicles racing, their paths through space for all times (worldlines) are shown as colored lines. Time is the vertical direction and space is the horizontal direction (units are omitted). They start the race at time $t = 0$ and spatial location $x = 0$, moving in the positive x direction. At $t = 5$, the car is at the finish line ($x = 15$) while the truck is only at $x = 5$. The truck does not get to the finish line until $t = 15$.

light. Conversely, Event B resides outside the future lightcone of Event A and is said to *spacelike related* to Event A, so Events A and B are said to be *spacelike separated*. This means a signal sent from Event A to Event B would be traveling faster than light. Causal influences between spacelike separated events are said to be *nonlocal* and Einstein derided them as "spooky actions at a distance." As we will see throughout, the word "causal" and the expression "causal influence" have various implications for physicists. Let us briefly unpack the relevant implications for the purposes of our discussion.

First, to say that "Event 1 causally influences Event 2" suggests that Event 1 causes Event 2 to come into being, i.e., Event 1 *produces* Event 2. Second, for Event 1 to produce Event 2 everyone must agree that Event 1 precedes Event 2 in time. Third, according to special relativity everyone will agree that Event 1 precedes Event 2 only if the relationship between Event 1 and Event 2 is timelike or lightlike, i.e., threads the lightcone (Figure 1.2). If the relationship between Event 1 and Event 2 is spacelike, then according to special relativity some observers will see Event 1 occur before Event 2 while others will see Event 2 occur before Event 1. This is why the lightcone structure of special relativity is called the *invariant causal structure* of the theory. All of this will become clear in future chapters.

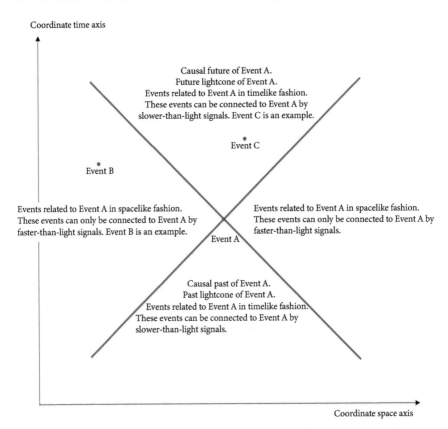

Coordinate time axis

Causal future of Event A.
Future lightcone of Event A.
Events related to Event A in timelike fashion.
These events can be connected to Event A by
slower-than-light signals. Event C is an example.

Event C

Event B

Events related to Event A in spacelike fashion.
These events can only be connected to Event A by
faster-than-light signals. Event B is an example.

Events related to Event A in spacelike fashion.
These events can only be connected to Event A by
faster-than-light signals.

Event A

Causal past of Event A.
Past lightcone of Event A.
Events related to Event A in timelike fashion.
These events can be connected to Event A by
slower-than-light signals.

Coordinate space axis

Fig. 1.2 A spacetime diagram where the units are such that light signals cross through Event A at 45°. Events along these diagonal lines are said to be *lightlike* related.

Fourth, since this relativistic causal structure is the result of Lorentz transformations[1] (Chapter 4), a relativistic causal mechanism must be Lorentz invariant, i.e., give the same predictions under Lorentz transformations. And finally, fifth, some physicists think a causal connection between any two events requires some contiguous causal medium such as a classical field. With all that in mind, here is why Einstein called nonlocal causal influences/connections "spooky actions at a distance."

Suppose Event A in Figure 1.2 is Alice sending a party invitation to Bob, and Event C is Bob reading the invitation. In that case, the invitation could be

---

[1] The Lorentz transformations simply convert the space and time coordinates of an event in one inertial reference frame to the space and time coordinates of that same event in a different inertial reference frame, such that all events are consistent with the light postulate.

traveling slower than light from Alice to Bob, so special relativity says everyone will agree that Event A occurred before Event C (this will be explained in Chapter 4). That is good because Bob cannot possibly read Alice's invitation (Event C) before she even sends it (Event A), that would certainly create confusion between cause and effect. This is why physicists highly value the invariant causal structure of special relativity. Now look at Events A and B.

Again, suppose Event A is Alice sending a party invitation to Bob, and Event B is Bob reading the invitation. In that case, the invitation was traveling faster than light from Alice to Bob, so special relativity says some observers will see Event B happen before Event A, while others will see Event B happen after Event A, and yet others will see Events A and B happen simultaneously. That means some observers would say Bob reads Alice's invitation before she even sent it. This is why Einstein derided spacelike causal relationships as "spooky" and why physicists are reluctant to invoke causal relationships between spacelike separated events (nonlocality). Now let's return to Maudlin's statement about the implications of entanglement.

In the case of entanglement, the events in question are the measurement events to include settings and outcomes. As Maudlin points out, the measurement settings are random, so presumably the quantum particles don't 'know' what measurements will be done on them until they arrive at their respective detectors (statistical independence). And, Alice and Bob's measurement events are spacelike separated, so any causal influence or signal between them would have to be nonlocal.

Again, Bell's theorem simply says that a causal mechanism acting in accord with locality and statistical independence must produce measurement outcomes that satisfy Bell's inequality (we will add two more assumptions, intersubjective agreement and unique experimental outcomes, in due course). Since we have "experimental verification of violations of Bell's inequality for randomly set measurements at space-like separation," we are faced with a challenge to "our fundamental account of the world," where "fundamental account" means "foundational account," no other account is more basic or unifying than the fundamental account.

Of course, it might be the case that the experimental verification of the quantum-mechanical predictions of the violation of Bell's inequality are all flawed for some reason. But, quantum mechanics has been experimentally verified in so many other respects that most physicists feel quite comfortable accepting the numerous experimental verifications of Bell inequality violations exactly as predicted by quantum mechanics (evidence the 2022 Nobel Prize in Physics). So, let us state the EPR–Bell paradox as:

- Bell's inequality is satisfied as long as certain reasonable assumptions are true.
- Quantum mechanics predicts the violation of Bell's inequality.

We need to be explicit as to why this is a paradox in order to resolve it.

Let's begin with a definition of paradox taken from Oxford Languages and Google:

a seemingly absurd or self-contradictory statement or proposition that when investigated or explained may prove to be well founded or true.

Now consider the example of a paradox from special relativity concerning the relativity of length contraction (alluded to in Chapter 0).

Given that Alice and Bob occupy inertial reference frames in motion relative to each other, the following two facts are true according to special relativity:

- Alice's measurements indicate that Bob's meter sticks are shorter than hers.
- Bob's measurements indicate that Alice's meter sticks are shorter than his.

If you are not familiar with the relativity of simultaneity from special relativity (which we will introduce in Chapter 4), these facts will probably strike you as "absurd or self-contradictory." Nonetheless, special relativity says they are both true. Assuming you have faith in the validity of special relativity otherwise, this set of facts creates a paradox for you, i.e., the set of (apparently true) facts is in conflict with your worldview (the facts seem to be self-contradictory). In order to resolve this paradox, we need to either correct the set of facts or revise your worldview. In this case, we will revise your worldview by adding the relativity of simultaneity from special relativity to your worldview and that will allow you to make sense of the facts, i.e., resolve the paradox.

In order for Alice to measure the length of Bob's meter stick, she needs to know where both ends of his meter stick are at the same time. Specifically, she needs to identify a pair of simultaneous events (two events in space that happen at the same time) whereby each event is at each end of his meter stick and then find the distance between those two events. The same is true for Bob's measurement of Alice's meter stick. According to the relativity of simultaneity, Alice and Bob simply disagree on where the ends of their meter sticks are at the same time because they disagree on which events are simultaneous. Again, for example, in Figure 1.2 some observers would say Event A happens before Event B (as shown), some will say Event B happens before Event A, and yet others will say Events A and B are simultaneous. Spacelike related events in the spacetime of special relativity have no definitive temporal order. Worldview updated, paradox resolved. Next, let's look at the EPR paradox.

The EPR facts can be stated as follows:

- Quantum mechanics is complete.
- There are no "spooky actions at a distance" (locality).

These two facts are "absurd or self-contradictory" because if quantum mechanics is complete, then no hidden variables can be used to account for the correlations of entangled particles when Alice and Bob are making the same measurements. In Einstein's own words [15]:

> By this way of looking at the matter, it becomes evident that the [EPR] paradox forces us to relinquish one of the following two assertions:
> (1) The description by means of the wave-function is complete.
> (2) The real states of spatially separated objects are independent of each other.

To see this, let's return to our classical gloves experiment of Chapter 0 as Jean Bricmont might state it [8].

In that experiment, when Alice measures the color of her glove she immediately knows what Bob will obtain for the outcome of a color measurement on his glove. That is not mysterious because the gloves have a color whether or not Alice and Bob choose to measure it, i.e., color is a hidden variable for the classical gloves.

But for the quantum gloves, assuming quantum mechanics does not need to be completed via hidden variables, Alice's measurement of quantum color must instantly 'create' the quantum color of that glove and, therefore, the quantum color of Bob's glove, however far apart they are. That means, without hidden variables (without completing quantum mechanics) we have only nonlocality ("spooky actions at a distance") to account for the correlations in violation of the second fact. So, our two EPR facts are contradictory according to our scientific worldview.

We should point out that EPR assumed statistical independence (introduced in Chapter 0 and above) and they only considered the case when Alice and Bob make the same measurement. In that case, hidden variables allow quantum mechanics to satisfy locality, but imply that quantum mechanics is incomplete. In deriving his inequality, Bell added the case when Alice and Bob make different measurements and discovered that hidden variables do not allow for locality in explaining the quantum-mechanical predictions (again, assuming statistical independence). Therefore, regardless of the completeness of quantum mechanics its predictions imply "spooky actions at a distance." Before revealing how we will resolve the EPR paradox, let us look more closely at Bell's extension thereof, i.e., the EPR–Bell paradox.

Clearly, we need to know those "certain reasonable assumptions" in our paradoxical EPR–Bell facts. That is, the EPR–Bell paradox is a paradox because quantum mechanics must violate at least one of Bell's "reasonable" assumptions and that must strike physicists and philosophers in foundations of quantum mechanics (the foundations community) as "absurd or self-contradictory," i.e., it must create a conflict with their scientific worldviews.

You can write the Bell assumptions like this [25][21]:

hidden variables + locality + statistical independence = Bell's inequality.

These correspond to Albert's three rules in Chapter 0. Accordingly, we start with the assumption that the entangled particles possess hidden variables responsible for the measurement outcomes. As Bell wrote [5]:

> In a complete physical theory of the type envisaged by Einstein, the hidden variables would have dynamical significance and laws of motion.

That is, the hidden variables are the cause of measurement outcomes and a complete theory of quantum physics would provide the dynamical details for that causation that are missing in quantum mechanics.[2] This is not something that can be easily abandoned in order to save locality and statistical independence. As Maudlin pointed out [28] and we explained above, EPR and Bell invoked hidden variables as necessary to save locality (given statistical independence), as will be clear in Chapter 2 with the Mermin device. If you abandon hidden variables, you are abandoning constructive explanation and you will need an entirely different type of explanation for violations of Bell's inequality, such as the principle explanation we will show. If you opt for a principle explanation, then locality and statistical independence are simply irrelevant, so you have not "saved locality" in that case either.

---

[2] Emily Adlam points out that it is possible for the hidden variables to not be hidden, e.g., the variables in spontaneous collapse models "are just given by the quantum state itself and the quantum state is accessible to observers" [3].

If you keep to constructive explanation, Bell's second assumption is that the hidden variables do not require information exchange between spacelike separated events (Figure 1.2) to do their work. Again, hidden variables are being invoked precisely to avoid "spooky actions at a distance." Finally, Bell's third assumption is that the measurement settings are random, so the hidden variables are not correlated with the measurement settings. Let's pause to clarify that last assumption.

In our quantum gloves experiment with Alice and Bob measuring only quantum color and quantum handedness, the hidden variables must be responsible for the Left and Right outcomes for quantum handedness measurements and the Black and White outcomes for quantum color measurements. Labeling the hidden variables according to the outcomes they produce we have four possible particles produced by the source: Left–White, Left–Black, Right–White, and Right–Black. As long as the source sends the same type of particle to Alice and Bob in each trial of the experiment, Alice and Bob will always get the same outcomes when making the same measurements without violating locality. Statistical independence just means that whatever rate the source produces each particle overall has to be the rate the source produces each particle for each of the four measurement choices.

For example, suppose the source produces Left–White particles 25% of the time, Left–Black particles 25% of the time, Right–White particles 25% of the time, and Right–Black particles 25% of the time. Then statistical independence says that the source produces that same distribution in the trials when Alice and Bob both measure quantum color, Alice and Bob both measure quantum handedness, Alice measures quantum color and Bob measures quantum handedness, or Alice measures quantum handedness and Bob measures quantum color. So, statistical independence fails if, say, the source never produces Left–White particles when Alice measures quantum color and Bob measures quantum handedness. How could something like that even happen? How could the source 'know' the eventual measurement settings before emitting the particles? If this is going on behind the scenes, measurement choices are not truly random in an insidious fashion, i.e., because the variables are hidden, Alice and Bob have no way to know that Nature is tricking them in this fashion. This will be made clear in Chapter 3 using the Mermin device of Chapter 2.

So, the paradoxical EPR–Bell facts are:

- Bell's inequality is satisfied as long as hidden variables, locality, and statistical independence are true.
- Quantum mechanics predicts the violation of Bell's inequality.

Again, EPR invoked hidden variables to save locality (and statistical independence) in constructive fashion, so we cannot give up hidden variables unless
we are willing to give up constructive explanation (in which case we are avoiding nonlocality, but not saving locality). Accordingly, quantum mechanics is
telling us that reality is nonlocal and/or it violates statistical independence if
we are wedded to constructive explanation. Bell writes [6, p. 149]:

> Let me summarize once again the logic that leads to the impasse. The EPRB
> correlations are such that the result of the experiment on one side immedi-
> ately foretells that on the other, whenever the analyzers happen to be parallel
> [Alice and Bob happen to make the same measurement]. If we do not accept
> the intervention on one side as a causal influence on the other, we seem
> obliged to admit that the results on both sides are determined in advance
> anyway, independently of the intervention on the other side, by signals from
> the source and by the local magnet setting [hidden variables]. But this has
> implications for nonparallel settings [for when Alice and Bob make differ-
> ent measurements] which conflict with those of quantum mechanics. So we
> *cannot* dismiss intervention on one side as a causal influence on the other.

Again, Bell is assuming statistical independence, therefore violations of his
inequality imply nonlocality if you believe reality is to be understood most fundamentally via constructive explanation. There is nothing "self-contradictory"
about that, but for those who demand a constructive account of entanglement,
the EPR–Bell paradox tells them they have to choose an explanation that is
nonlocal and/or violates statistical independence. Simply put, most members
of the foundations community find that to be "absurd." Let's look briefly at
why nonlocality and violations of statistical independence are objectionable
to so many in the foundations community. We will also introduce violations
of intersubjective agreement and unique experimental outcomes as ways to
explain the correlations of entanglement and show why those are also not
popular.

1.2 Responses to the EPR and EPR–Bell Paradoxes

If you employ faster-than-light causal influences ("spooky actions at a distance"), you can "restore to the theory causality," but at the cost of locality. Since special relativity says the temporal order of spacelike related
events depends on your reference frame, the causal ordering of such events

is ambiguous without a preferred reference frame (Figure 1.2). Therefore, invoking spacelike causation (giving up locality) in order to save causality means invoking a preferred reference frame, which is at odds with the relativity principle and the invariant causal structure of special relativity, both of which are highly regarded in the foundations community.

If you give up statistical independence by invoking backward-in-time causation (called *retrocausality*), whereby particle properties at emission are determined by future measurement choices, you do "restore to the theory ... locality." However, this is at the cost of causality because you lose what many consider essential to causal explanation, i.e., causes precede their effects in time. One version of this is called *supermeasured theory*, where "measured" refers to a probability measure, not measurement (Chapter 3).

If you give up statistical independence by proposing 'conspiratorial' causation whereby events in the distant past are causally responsible for both the particle properties *and measurement settings* (called *superdeterminism*), you can "restore to the theory causality and locality." However, this is at the cost of what most consider essential to experimental physics, i.e., the ability to make truly random measurements as described above.

Besides violating locality and/or statistical independence, there are two other accounts of quantum mechanics that we will highlight due to their (relative) prominence. We will not always include these additional accounts when discussing Bell's assumptions specifically, but we will refer to them when discussing accounts of quantum mechanics for dealing with entanglement generally. Let us introduce those now.

If you give up the fact that there is only one measurement outcome for an experiment (as in the Many-Worlds interpretation), you can "restore to the theory causality and locality," but you introduce what is called the *probability problem*. That is, since all possible outcomes occur (each in its own distinct 'world'), you need some rule for locating those versions of 'you' through the near infinite maze of your possible 'selves' that measures the outcomes at the rate given by the quantum state.

For example, suppose you are measuring quantum color for a quantum glove in each trial of the experiment and the quantum state is 25% White and 75% Black (rather than 50% White and 50% Black as in the Chapter 0 example). You will split into two 'selves' each time each 'you' conducts the experiment (because there are two possible outcomes), but the correct rate of White and Black outcomes is $\frac{1}{4}$ White and $\frac{3}{4}$ Black rather than $\frac{1}{2}$ White and $\frac{1}{2}$ Black. So, any given 'you' must proceed through the branching worlds such that they encounter three Black outcomes for every White outcome. Of course, if you simply try to interpret Many-Worlds without probability, then you run into a problem with empirical confirmation [1], so that will not work either. The question is, what makes any given 'you' proceed through the branching worlds in accord with the empirically verified probabilities of quantum mechanics?

Additionally, while this removes the problem of nonlocality in any given world, nonlocality still exists between worlds in the larger sense [43]. That is because the splitting into worlds only produces the outcome pairs given by the restricted, entangled quantum state to begin with. So when Alice and Bob happen to measure the same spin for the entangled quantum gloves experiment, there are only two possible outcomes, Alice gets +1 and Bob gets +1 ([+1, +1]) or Alice gets –1 and Bob gets –1 ([–1, –1]), rather than all four possible combinations of +1 and –1 ([+1, +1], [+1, –1], [–1, –1], [–1, +1]) that occur when Alice and Bob happen to measure different spins. The world that is splitting in accord with just two outcomes or all four outcomes does so even if Alice and Bob are light-years apart in that world.

The bottom line is, while Many-Worlds removes nonlocality for any given 'you', it does not provide a solution to the mystery of entanglement because entanglement is fundamental in Many-Worlds. That is, entanglement is used to explain or account for aspects of Many-Worlds, but entanglement itself is not explained in terms of anything more foundational or basic. This emphasis on time evolution of the wavefunction makes Many-Worlds a dynamical (and therefore constructive) account of quantum mechanics and the constructive bias driving Many-Worlds will be on display most prominently when we explain Many-Worlds as a response to the measurement problem in Chapter 3.

Finally, if you give up trying to reconcile all the different experiences of all the different observers self-consistently into a single, intersubjectively-agreed-upon (objective) model of reality, you can "restore to the theory causality and locality," but arguably one is no longer doing science. Indeed, one can argue that it is contradictory to interpret a theory constructed according to inter-subjective agreement by positing the violation of intersubjective agreement [2]. Certainly, most physicists believe they must reconcile their experimental outcomes with the experimental outcomes of their colleagues when doing the same experiment via a commonly held objective model of reality. Zeilinger writes [47]:

We have knowledge, i.e., information, of an object only through observation. Thus, any concept of an existing reality has to be based on observations. Yet this does not imply—as tempting as such a conclusion might be—that reality is no more than a pure subjective human construct. From our observations we might mentally construct objects of reality. Predictions based on any such specific model of reality may then be checked by anyone. As a result we may arrive at intersubjective agreement on the model, thus lending a sense of objectivity to the mentally constructed objects.

Abandoning a unified (objective) model of reality ("intersubjective agreement on the model") rids us of nonlocality because each observer's experiences are timelike related along their worldline, so a person's individual experiences are local for them by definition. Essentially, you are abandoning the concern over spacelike correlations because you are abandoning any meaning to a common spatial distribution of entangled events.

The particular version of this proposed solution to the mystery of entanglement that we will discuss is an older version of Carlo Rovelli's relational quantum mechanics [35]. As we will discuss in Chapter 9, Adlam and Rovelli have since altered relational quantum mechanics so that it conforms to intersubjective agreement [4], but relational quantum mechanics is still of particular interest to us because Rovelli bases his view of quantum mechanics on information theory. More precisely, he posits the same information-theoretic principle at the foundation of quantum mechanics that we will use, i.e., Information Invariance & Continuity per Zeilinger and Časlav Brukner [9], albeit in a different form. For example, Philipp Höhn's reconstruction "has been inspired by Rovelli's relational quantum mechanics and the Brukner–Zeilinger informational interpretation" [24]. These accounts of quantum mechanics are not necessarily constructive, e.g., we would say relational quantum mechanics is a principle account.

We will detail each of these approaches in Chapter 3, but from the brief overview given here you can appreciate why none of these options has won consensus support in the foundations community. The cost of violating locality, statistical independence, intersubjective agreement, or unique experimental outcomes is just deemed too high for some majority of the foundations community. So, what exactly is our resolution of the EPR and EPR–Bell paradoxes?

1.3 How Will We Resolve These Paradoxes?

Given that the Bell inequality is mathematically true and quantum mechanics is a highly successful theory applicable where it predicts the violation of Bell's inequality, we are going to propose revising the scientific worldview with its constructive bias so that it no longer conflicts with the EPR and EPR–Bell facts. And, in what will prove to be possibly the greatest irony in physics, we are going to follow Einstein's lead and solve the mystery of entanglement, as necessary to resolve our paradoxes, just as he solved the mystery of length contraction. That is, we are going to abandon constructive explanation and turn instead to principle explanation à la Einstein's distinction between constructive and principle theories.

Essentially, we are proposing that principle explanation is sometimes the most foundational explanation, not constructive explanation. According to this change in our constructive scientific worldview, it is quite reasonable that some phenomena can only be explained in universally acceptable fashion via principle explanation. So, what exactly do we mean by principle and constructive explanation?

According to Einstein, constructive theories are based on dynamical laws and/or mechanistic causal processes. He used the kinetic theory of gases as an example of a constructive theory [12]. Accordingly, properties such as the temperature and pressure of the gas are understood by averaging dynamical and causal facts for the particles of the gas. Temperature is associated with the average kinetic energy of the gas particles and pressure is associated with the average change in momentum of the gas particles in collisions with the container walls.

By analogy, let us define constructive explanation as explanation based on dynamical laws and/or mechanistic causal processes (causal mechanisms). Again, Bell's assumptions (and Many-Worlds) are based on constructive explanation. Constructive efforts have all failed to resolve the EPR and EPR–Bell paradoxes, i.e., to provide a consensus solution to the mystery of entanglement, because they all require a violation of locality, statistical independence, intersubjective agreement, or unique experimental outcomes. So, we will use a principle explanation to solve the mystery of entanglement in analogy with Einstein's definition of a principle theory.

According to Einstein, principle theories are based on an empirically discovered fact that constitutes our fundamental principle [12], i.e., the principle at the foundation of the explanatory hierarchy. The mathematical consequences of that fundamental principle then create the principle theory. Einstein used thermodynamics as an example of a principle theory. The fundamental principle / empirically discovered fact at its foundation is "perpetual motion machines are impossible." The laws of thermodynamics follow accordingly, i.e., perpetual motion machines of the first/second/third kind violate the first/second/third law of thermodynamics. Of course, special relativity is also a principle theory (as well as a principle explanation), as we will explain below.

Again, we shall define principle explanation in analogy with Einstein's definition of a principle theory. We will deviate slightly from his definition in that we will make a distinction between the compelling fundamental principle and the empirically discovered fact at the foundation of a principle explanation.

Another way to view the distinction we are trying to make here was proposed by Diego Maltrana, Manuel Herrera, and Federico Benitez [26]:

> Those theories that allow us to *trace* the *causal mechanisms* that *explain mechanistically* the occurrence of a certain phenomenon we call "mechanistic theories." And those theories that lack agents whose actions are causally responsible for phenomena, but that instead provide general constraints or structural elements that lead to unificationist explanations we call "structural theories."

Accordingly, a mechanistic theory, e.g., electromagnetism, provides explanations based on causal mechanisms, e.g., the Lorentz force law, while a structural theory, e.g., special relativity, provides explanations based on general constraints, e.g., the invariance of the speed of light, or structural elements, e.g., spacetime geometry per the Lorentz transformations (geometry of Minkowski spacetime). Obviously, their mechanistic theory corresponds to Einstein's constructive theory and their structural theory corresponds to Einstein's principle theory.

It is important to note that some scholars, e.g., Laura Felline [17], use the term "principle explanation" synonymously with the term "structural explanation," but we will distinguish between structural explanation per Felline [17,18,19,20] and our principle explanation [40]. Specifically, the most fundamental explanatory element (explanans) for principle explanation is a compelling fundamental principle to justify the empirically discovered fact at the foundation of a principle theory, while the explanans for structural explanation per Felline is the formal structure of the principle theory itself.

For special relativity as a principle explanation, the explanans is the relativity principle, since it justifies the light postulate. And for special relativity as a structural explanation, the explanans is the geometry of Minkowski spacetime. For example, length contraction follows most fundamentally from the relativity principle according to principle explanation, while it follows most fundamentally from Minkowski spacetime geometry according to structural explanation. This will prove to be an important distinction when considering the kind of explanation provided by the information-theoretic reconstructions of quantum mechanics (Chapter 6).

Here is the form of a principle explanation of / solution to some mystery:

compelling fundamental $\rightarrow$ justifies empirically $\rightarrow$ dictating the
principle   discovered fact   mystery

This is what Einstein turned to out of 'despair' at the beginning of the 20th century when physicists (himself included) were considering increasingly complex constructive explanations of length contraction (Chapter 4).

In 1865, James Clerk Maxwell published equations of electromagnetism that predicted the existence of electromagnetic waves moving at the speed of light c, so light was accepted as an electromagnetic wave. The question was, what was waving? Physicists called the hypothetical waving medium the *luminiferous aether* and set about finding Earth's speed through it. Albert Michelson and Edward Morley developed a very accurate means of measuring this speed and to everyone's surprise they found it to be ... zero.

In desperation, George FitzGerald and Hendrik Lorentz conjectured that an object's length would shrink along the direction of its motion in the aether and this length contraction would then explain the Michelson–Morley null result. The problem then became to provide a constructive explanation for this length contraction. That turned out to be much more difficult than expected and every attempt to explain length contraction constructively failed to win consensus support. Sound familiar?

Even Einstein tried and failed to explain the Michelson–Morley null result constructively and gave up, saying [14]:

I despaired of the possibility of discovering the true laws by means of constructive efforts based on known facts.

So, he flipped the explanatory hierarchy by turning to a principle explanation. That is, after considerable effort, he gave up in despair of finding a constructive explanation for length contraction to explain the Michelson–Morley null result via causal mechanisms. The Michelson–Morley null result can be stated as the light postulate, i.e., everyone measures the same value for c, regardless of their relative motions. In other words, he gave up trying to do this:

aether $\rightarrow$ length contraction $\rightarrow$ light postulate.

Instead, he turned to principle explanation and invoked the relativity principle to justify the light postulate (empirically discovered fact), then used *that* to derive length contraction (Chapter 4):

$$\text{relativity principle} \rightarrow \text{light postulate} \rightarrow \text{length contraction.}$$

Since the relativity principle says the laws of physics (including their constants of Nature) are the same in all inertial reference frames, we will refer to it as "no preferred reference frame" (NPRF). Reference frames in uniform motion with respect to each other constitute inertial reference frames and c is a fundamental constant of Nature per Maxwell's equations, so Einstein proposed NPRF as the compelling fundamental principle to justify the light postulate. As a result, we understand that the principle account of length contraction ultimately resides in the relativity principle and we will say that the mystery of length contraction was solved in principle fashion by NPRF + c.

In Chapter 5, we contrast principle and constructive explanation and discuss the possibility that principle explanation is really fundamental to constructive explanation with regard to certain phenomena. That is to say, the relativity principle is more fundamental than Reichenbach's Common Cause Principle (Reichenbach's Principle), which states:

If events A and B are correlated, then A causes B, B causes A, or there is some common cause C for both A and B in their past.

Reichenbach's Principle is historically significant because if it is true, it licenses causal inference from probabilistic correlations. We will argue that the correlations exhibited by quantum entanglement are not best understood in terms of Reichenbach's Principle or any variant of it [36]. That is, entanglement is best understood as an "all-at-once" or "atemporal" constraint on the distribution of quantum events in spacetime. We will also refer to this as an *adynamical* or *acausal global constraint* (AGC).

Given that Einstein's principle is more fundamental than Reichenbach's, and given the historical precedent of special relativity, we will resolve the EPR and EPR–Bell paradoxes in principle fashion. In what might be the greatest irony in physics, we provide a possible solution to "the greatest mystery in physics" with the relativity principle. That is, Einstein's "spooky actions at a distance" can be dispelled by Einstein's own relativity principle (Chapter 7). To understand how that happens, we first take a brief excursion through quantum information theory and its axiomatic reconstructions of quantum mechanics (Chapter 6).

More specifically, we will show that the invariant value of another fundamental constant of Nature, i.e., Planck's constant h, between inertial reference frames related by spatial rotations (Planck postulate) dictates the mystery of entanglement, i.e., 'average-only' conservation [39, 41, 42]. Planck introduced the constant h as part of his blackbody radiation equation and Einstein used it for his photoelectric effect equation. Both of these rely on the Planck–Einstein relation, $E = hf$, where E is the energy of an electromagnetic mode (Planck) or photon (Einstein) with frequency f. Accordingly, we will say that NPRF + h solves the mystery of entanglement in principle fashion just as NPRF + c solves the mystery of length contraction in principle fashion. In other words, in total analogy with special relativity:

relativity principle → Planck postulate → 'average-only' conservation.

That means NPRF + h resolves the EPR paradox because quantum mechanics is as complete as possible given NPRF + h (first EPR fact), and there is no need to invoke "spooky actions at a distance" to solve the mystery of entanglement (second EPR fact). So, changing the scientific worldview by accepting principle explanation in lieu of constructive explanation for entanglement means the two EPR facts are both true without contradiction.

NPRF + h also resolves the EPR–Bell paradox because both of the paradoxical facts are true without absurdity if you solve the mystery of entanglement in principle fashion. Again, the EPR–Bell facts are only absurd if you require a constructive explanation of the second EPR–Bell fact (quantum mechanics predicts the violation of Bell's inequality) because in that case the first EPR–Bell fact (Bell's inequality is satisfied as long as hidden variables, locality, and statistical independence are true) says your constructive explanation must violate locality and/or statistical independence, which constitutes an unacceptable (absurd) explanation. Indeed, some would say the violation of locality or statistical independence (or intersubjective agreement or unique experimental outcomes) contradicts what they deem *necessary* for a proper constructive explanation to begin with.

By using a principle (rather than constructive) explanation of the second EPR–Bell fact, the validity of the first EPR–Bell fact is irrelevant, which removes the absurdity. Since we must change the scientific worldview to allow for *some* explanation of entanglement that is not constructive in order to resolve the paradox, why not a principle explanation that closely mirrors that of special relativity? And, the relativity principle goes all the way back to Galileo (and further [30], but we'll only go back to Galileo).

1.4 A Brief History of NPRF

Before Einstein invoked the relativity principle to replace the Newtonian view of space and time with spacetime in 1905 (Figure 1.2), Galileo invoked it to challenge Aristotle's 'teleological' view of reality in 1632. That led to Isaac Newton's 'mechanistic' view of reality in his 1687 publication *Principia* [10, 11]. John Norton writes [31]:

> While not present by name, the principle of relativity has always been an essential part of Newtonian physics. According to Copernican cosmology, the earth spins on its axis and orbits the sun. Somehow Newtonian physics must answer the ancient objection that such motions should be revealed in ordinary experience if they are real. Yet, absent astronomical observations, there is no evidence of this motion. All processes on earth proceed just as if the earth were at rest. That lack of evidence, the Newtonian answers, is just what is expected. The earth's motions are inertial to very good approximation; the curvature of the trajectory of a spot on the earth's surface is small, requiring 12 hours to reverse its direction. So, by the conformity of Newtonian mechanics to the principle of relativity, we know that all mechanical processes on the moving earth will proceed just as if the earth were at rest.

According to Aristotle, the world is divided into two realms, the Earthly realm and the Heavenly realm. Corrupt and changeable objects (made of earth, air, fire, and/or water) are pulled by gravity towards the center of the universe (Earth) because that's where they 'belong'. Levity lifts perfect and immutable objects (made of aether, also known as quintessence, for the fifth element) to the Heavenly realm where they 'belong' and execute uniform circular motion about Earth (the center of the universe), since the circle was regarded as the most perfect geometric form.

The natural state of Earthly objects was that of rest, so an Earthly object required a 'mover' to keep it moving. People believed this because it is empirically verified, as you can surely attest from your own personal experience. For example, if you throw an object, it comes to rest. Nothing in our everyday experience keeps moving unless it is made to move. That's why it was ridiculous to believe Earth orbited the Sun. What mover would keep the entire Earth in motion around the Sun?

But, there was an empirical challenge to the Aristotelian worldview, i.e., some Heavenly objects did not seem to execute uniform circular motion about Earth. These objects were called *planets*, which means 'wanderers,' because they moved relative to the stars in addition to their motion relative to Earth. Indeed, the planets sometimes orbited Earth in the *opposite* direction to the stars. Additionally, planets did not maintain the same distance from Earth, their distance changing between perigee (nearest distance from Earth) and apogee (farthest distance from Earth).

Claudius Ptolemy explained this retrograde motion and varying distance by having the planets execute uniform circular motion about points in space (epicycles) that were themselves executing uniform circular motion about Earth (deferents). In addition to epicycles on deferents, Ptolemy had to displace Earth a bit from the center of the universe (equant) in order to fit all the observations (Figure 1.3). This same trend towards increasingly complex constructive explanations was evident in constructive models of the late 19th century used to account for length contraction, and it is being played out again today to account for entanglement.

Just like Einstein used the relativity principle to challenge increasingly complex constructive explanations of length contraction, Galileo invoked the relativity principle to challenge increasingly complex constructive explanations of planetary observations. These observations were interpreted via the premise that Earth was immobile and located at the center of the universe,

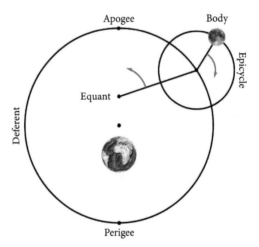

Fig. 1.3 The Ptolemaic model of astronomy. In order to account for the retrograde motion of the planets while having everything orbit Earth and execute uniform circular motion, Ptolemy's model became ever more complex. Instead of simply orbiting Earth in uniform circular motion like the stars, planets orbited in epicycles on deferents with an equant.

since Earth needed a mover to orbit the Sun. Galileo invoked the relativity principle to challenge that premise in 1632 [22, pp. 186–187]:

> Shut yourself up with some friend in the main cabin below decks on some large ship, and have with you there some flies, butterflies, and other small flying animals. Have a large bowl of water with some fish in it; hang up a bottle that empties drop by drop into a wide vessel beneath it. With the ship standing still, observe carefully how the little animals fly with equal speed to all sides of the cabin. The fish swim indifferently in all directions; the drops fall into the vessel beneath; and in throwing something to your friend, you need throw it no more strongly in one direction than another, the distances being equal; jumping with your feet together, you pass equal spaces in every direction. When you have observed all these things carefully (though there is no doubt when the ship is standing still everything must happen in this way), have the ship proceed with any speed you like, so long as the motion is uniform and not fluctuating this way and that. You will discover not the least change in all the effects named, nor could you tell from any of them whether the ship was moving or standing still.

That is, there is nothing special about the rest frame of Earth. Newton recognized this and included Galileo's story in the *Principia* [30]. Accordingly, Newton's first law of motion, "an object in motion tends to remain in motion and an object at rest tends to remain at rest," differs significantly from Aristotle's law of motion, "an Earthly object in motion comes to rest and an Earthly object at rest remains at rest." First, for Newton there is no difference between Earthly and Heavenly objects, there are just objects. Second, as Galileo pointed out, there is nothing special about being at rest with respect to Earth. The reason your everyday objects come to rest after an initial thrust is because of the ubiquitous force of friction. Aristotle's worldview was constructive, but there was a teleological reason for its forces while Newtonian forces acted without purpose.

So, while much generality and unification were gained in explanatory power by moving from Aristotelian physics to Newtonian physics, an element of explanation was lost: a purpose behind the forces. In Newtonian mechanics, one simply identifies the mathematical form and context for forces then uses them where applicable. Why those forces act as they do is not a concern for Newton's physics. In Aristotle's physics, gravity or levity acted on objects because those objects 'belong' in either the Earthly realm or Heavenly realm, respectively. We will see this trade-off between increased generality in explanatory power and decreased explanatory specificity continue with each subsequent implementation of the relativity principle.

For example, Einstein generalized Galileo's version of the relativity principle, "The laws of mechanics are the same in all inertial reference frames," to "the laws of physics (mechanics and electromagnetism) are the same in all inertial reference frames." Since the speed of light c is part of Maxwell's equations for light, Einstein's version of the relativity principle demanded observers in all inertial reference frames measure the same value for c. That gives you the kinematics (Lorentz transformations) for special relativity with its length contraction.

Thus, Einstein's more general version of the relativity principle unified mechanics and electromagnetism, showing us that Newtonian mechanics is only a low-speed approximation to special relativistic mechanics. However, as with the transition from Aristotle's physics to Newton's physics, the transition from Newton's physics to Einstein's physics comes with a trade-off, i.e., a fundamental causal mechanism enforcing NPRF + c is not a concern for Einstein's physics. All of this leads to the marriage of relative space and relative time into a spacetime for the Lorentz transformations with its invariant causal structure (explained above).

That is, while not everyone agrees on the length or the time between events in Minkowski spacetime, they all do agree on the spatiotemporal distance between those events. And, observers in all reference frames also agree on the causal ordering of events in Minkowski spacetime, as we explained above. So, while special relativity is a theory based on the relativity principle as opposed to dynamical laws and/or mechanistic causal processes, it does cohere with causal explanation. In contrast, the implementation of the relativity principle being introduced here to solve the mystery of entanglement brings about a much more serious threat to the fundamentality of constructive explanation (Chapter 9).

1.5 Why is NPRF so Compelling?

In order to appreciate why the relativity principle is a compelling fundamental principle of physics, we need to understand its relevance to the very practice of physics itself. And to do that, we need to understand the practice of physics down to its very essence. No one articulated that better than Einstein with his 1936 essay "Physics and Reality" [13].

1.5.1 First and Second Points Per Einstein

According to Einstein [13], "The whole of science is nothing more than a refinement of everyday thinking" so the "game" (practice) of physics begins with some methodological and phenomenological facts, i.e., facts about "sense experiences" that apply to everyone doing physics. In other words, he simply stated the obvious. Now it may seem ridiculous to articulate the practice of physics all the way down to such a trivial level, but this is precisely where the relativity principle plays a role, so we cannot ignore these details.

We begin with the fact that [13]

> ... physics treats directly only of sense experiences and "understanding" of their connection. But even the concept of the "real external world" of everyday thinking rests exclusively on sense impressions.

With this point, Einstein is simply stating the obvious fact that physics deals most fundamentally with our observations, i.e., physics is an empirical science. Acknowledging this trivial phenomenological fact is crucial, as we shall see. [Note: In Chapter 9 we will explain why this does not entail operationalism or instrumentalism.[3]]

Second, Einstein points out that his observations involve interacting "bodily objects" like balls, trees, cars, etc. that have worldlines (or worldtubes if you take into account their spatial extent) in a spacetime diagram. He is also assuming that everyone else is making similar observations ("partly in conjunction with sense impressions which are interpreted as signs for sense experiences for others"). So, each person makes observations personally and with data collection devices that they use to create a personal (subjective) spacetime model of reality for the interacting bodily objects of their empirical data. Before proceeding to Einstein's third point, let's pause to establish some facts about these personal (subjective) spacetime models of reality.

1.5.2 Subjective Spacetime Models of Reality

We begin with the fact that bodily objects have worldlines/worldtubes in spacetime but they only exist instant by instant in our experience. For

---

[3] According to operationalism, scientific concepts only have meaning in terms of how their formalism operates. According to instrumentalism, scientific theories do not necessarily tell us anything about reality, they are simply an instrument for predicting the outcomes of experiments.

example, suppose the worldline for a probe heading to Jupiter is being created electronically from data transmitted from the probe to you on Earth via a radio signal. Since radio waves travel at the speed of light and your probe is almost at Jupiter, the radio signals take about 45 minutes to get from the probe to your data analyzer on Earth. So, the spatial and temporal location of the probe on the digitally created spacetime diagram is not where the probe is now, but where it was 45 minutes ago. Those versions of the probe and you no longer exist, only you now and the probe now exist. To emphasize the importance of this fact, we will describe a spatial slice of your subjective spacetime model of reality as a NOW slice. This NOW slice contains all coexisting (simultaneously existing) bodily objects and events in your subjective spacetime model of reality.

Although you are not sure where the probe is on your current NOW slice, you can use all of the past trajectory data and Newton's laws of motion to locate the probe in Space NOW (Figure 1.4). You need to do that because you need

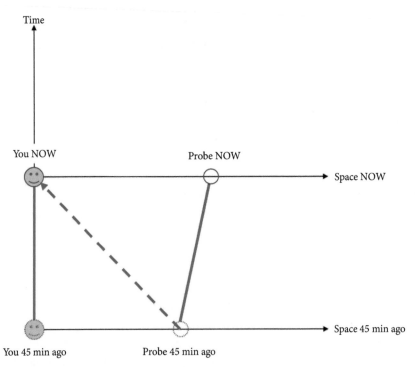

Fig. 1.4 Your personal (subjective) spacetime model of you on Earth and a probe collecting data for you en route to Jupiter. The worldlines for you and the probe in spacetime are in red. The radio signal from the probe to you is represented by the blue dashed arrow at 45° as in Minkowski spacetime (Figure 1.2).

to send navigation commands to the probe based on where it currently exists, since you will use that location to predict where it will be in 45 minutes when it receives your navigation commands.

While it is not done explicitly, everyone uses a personal spacetime model of reality every day; again, "the whole of science is nothing more than a refinement of everyday thinking." Typically the speed of our signals is fast enough and exchanged between locations in Space that are close enough together that we do not have to worry about Time delays like that of the Jupiter probe.

For example, suppose you send a text message to your friend asking them where they are and they tell you they are at the beach. Unlike the probe, your friend's response reaches you in mere seconds, so you go ahead and map (in your mind) your friend's location at the beach NOW. That's where they exist in the Space NOW of your subjective spacetime model of reality. You head to the beach to be with your friend and imagine you will arrive there in 45 minutes, i.e., you place yourself with your friend in Space at the beach 45 minutes into the future in your subjective spacetime model of reality (Figure 1.5).

Of course, you do not literally draw a spacetime diagram, but essentially that is what you are doing in your mind when you think about the locations of bodily objects and yourself in Space NOW relative to their locations in the past or in the future. Again, it may seem silly to think about everyday experience in terms of something as technical as a subjective spacetime model of reality, but you will appreciate why we are making this explicit later.

To summarize, we understand the following intuitive facts in terms of our everyday, functional, subjective spacetime models of reality (some of which will be challenged later):

- We and the bodily objects of our experience / empirical data exist in Space and persist through Time. These facts are represented by world-lines/worldtubes in spacetime.
- We and the bodily objects of our experience / empirical data only exist NOW. That is, the versions of them in past Space slices no longer exist and the versions of them in future Space slices are yet to exist.
- Information-carrying signals exchanged by these bodily objects can cause them to alter their behavior in Space as a function of Time according to the relevant dynamics.
- The information-carrying signals between bodily objects move through Space at a maximum speed of 300 000 km/s, i.e., the speed of light c. This

speed is so large that it is assumed to be infinite for most everyday contexts, i.e., for all practical purposes we often assume $c = \infty$ as in Newton's spacetime.

- The information-carrying signals between bodily objects are represented by worldlines in spacetime.

Armed with this intuitive understanding of our subjective spacetime models of reality, we are ready to return to Einstein's third point in the practice of physics.

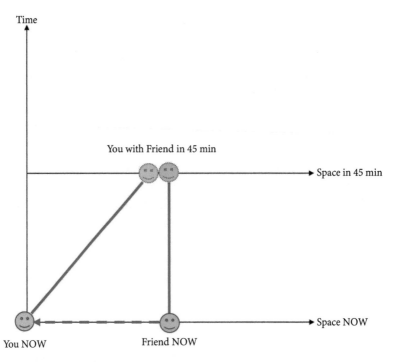

Fig. 1.5 Spacetime diagram of you and a friend coexisting in Space NOW. The worldlines for you and your friend in spacetime are in red and the text message from your friend at the beach to you at your house is represented by the blue dashed arrow. Since that information reaches you in mere seconds, it is mapped as instantaneous on the diagram, which effectively means $c = \infty$ as in Newton's spacetime. The subjective spacetime model of reality shows your intention to leave your house and drive 45 minutes to be with your friend at the beach, i.e., you expect to coexist with your friend at that location in Space in 45 minutes. Your worldline is 40° from the Time axis, which would be nearly the speed of light in a Minkowski spacetime diagram (Figure 1.2).

1.5.3 Einstein's Third Point and NPRF

Recall Zeilinger's statement from above:

> Predictions based on any such specific model of reality may then be checked by anyone. As a result we may arrive at intersubjective agreement on the model, thus lending a sense of objectivity to the mentally constructed objects.

This refers to Einstein's third point in the practice of physics. That is, working collectively with their joint data, physicists are trying to discover patterns in the relationships between, and events involving, the bodily objects of their empirical investigations. Einstein writes [13]:

> ... the totality of our sense experiences is such that by means of thinking (operations with concepts, and the creation and use of definite functional relations between them, and the coordination of sense experiences to these concepts) it can be put in order ...

Specifically, physicists seek to codify these patterns in precise mathematical expressions, which we then call *laws of physics*. As Mermin puts it [29]:

> Laws of science are the regularities we have discerned in our individual experiences, and agreed on as a result of our communications with each other.

So, the game of physics deals fundamentally with observers who collect, exchange, and coherently synthesize the information from their individual data collection devices into an objective spacetime model of reality. That is, they create an objective spacetime model of reality from their individual subjective spacetime models of reality. Everyone can then use the objective spacetime model of reality (arrived at by intersubjective agreement) to understand/explain why events in their (or anyone else's) personal past happened, and they can predict events in their (or anyone else's) personal future.

The Time axis of the objective spacetime model of reality can be the worldline of any inertial (non-accelerated) bodily object (observed or imagined), regardless of its state of motion, spatial orientation, or spatial position relative to the other bodily objects with worldlines in the model. Each such choice is called an *inertial reference frame* [34]. Given this meticulous explanation of the practice of physics, we see immediately why physicists find the relativity principle to be so compelling.

As long as the data collection device is inertial, we have no reason to treat its data differently from the data of any other data collection device in any other inertial reference frame. However, the world might be such that the data from different inertial reference frames does not conform to the same laws of physics, i.e., some law or constant might only take its 'correct' form or value in a particular inertial reference frame.

Indeed, when physicists realized that Maxwell's equations predicted a particular value for c, they immediately started looking for the (preferred) reference frame in which one would measure that particular value. When all inertial reference frames started producing that value (light postulate), they were perplexed and began to work on increasingly complicated constructive accounts of how Nature (or God or the Deity) was fooling them. That is, movement through the preferred inertial rest frame of the aether resulted in length contraction in just such a way that everyone measured the same value for c, regardless of their inertial reference frame. Of course, we explained above how Einstein resolved that.

Likewise, as we will see in Chapters 6 and 7, given the Planck–Einstein relation (from 1900), Einstein could have used the relativity principle to justify the Planck postulate (from the 1922 Stern–Gerlach experiment, see Chapter 2) and derive the (finite-dimensional) Hilbert space of quantum mechanics in the 1920s, just as the relativity principle justifies the light postulate, whence Minkowski spacetime. Had he done so, the EPR paradox would never have been born in 1935 because it would have been clear that the mystery of entanglement is simply the result of NPRF, just like the mystery of length contraction. No EPR paradox, no EPR–Bell paradox, and no violation of locality, statistical independence, intersubjective agreement, or unique experimental outcomes. But, Einstein did not do that and we now have the quantum-mechanical Hilbert space in widespread use, so we are invoking Einstein posthumously and using NPRF to help the foundations community come to terms with the crazy consequences of quantum mechanics, e.g., entanglement.

1.6 "God's Thoughts When He Created the World"

In Chapters 0–8 we focus on phenomenology and methodology for constructing an objective spacetime model of reality. We solve the mystery of entanglement this way so as to invoke minimal ontological[4] assumptions,

---

[4] Ontology is the study of what exists physically, e.g., electrons, photons, cars, people, etc. It must contain some constructive elements if it is to map onto our dynamical experience, but theories and explanations in that constructive ontology can be principle or constructive, as explained above.

which is easy to do for a principle explanation by its very nature. This way, readers can choose their own ontologies consistent with our principle explanation if they like. However, that means initially we do partly dodge Zeilinger's question, "What's really going on?"

So, in Chapter 9 we allow ourselves to fully speculate on the fundamental explanatory principles and the entities they range over, i.e., Einstein's desire to know "God's thoughts when He created the world" [32]. We will make it clear in Chapter 9 that far from entailing instrumentalism or operationalism, our principle explanation of quantum mechanics will allow us to simply accept the basic ontology already given by physics, such as relativistic and quantum particles and fields. There is no need to create or invoke an additional "primitive" physical quantum ontology underneath what is already given. The next subsection spells this out in a little more detail and suggests how we can begin to simplify all of this in terms of the quantum "bit" of information (qubit).

1.6.1 Two Theories, Two Constants, and Four Experiments

The ontology we propose is based on the Poincaré symmetry group, i.e., invariance under relativistic boosts, spatial rotations, spatial translations, and temporal translations (Figure 1.6). This will come as no surprise to particle physicists, who build their quantum field theories around Poincaré symmetry most fundamentally. We do not otherwise discuss quantum field theory, but we will very briefly and heuristically reference it for this topic in Chapter 9. While the necessary special relativity and quantum mechanics will take many pages of the book to fully unpack for making this point, let us hint at it now.

Historically, Maxwell's equations gave us the constant of Nature c, and the relativity principle dictates that c must have the same value in all inertial reference frames, even those in relative motion (light postulate) as discovered in the Michelson–Morley experiment. This gives us special relativity (Chapter 4). Planck's blackbody radiation equation gave us the constant of Nature h, and the relativity principle dictates that h must have the same value in all inertial reference frames (Planck postulate). We will share three experiments in which the invariance of h between inertial reference frames sheds light on the mathematical structure of quantum mechanics per quantum information theory, and all of these bear on the fact that h is a fundamental unit of action (Figure 1.6).

Action is angular momentum multiplied by angular distance, or energy multiplied by time, or (linear) momentum multiplied by distance, and then added up along the worldline for a particle. All theories of physics can be cast in

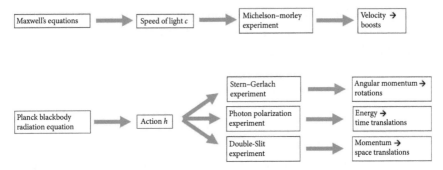

Fig. 1.6 Two theories, two constants and four experiments for a Poincaré symmetric ontology.

terms of action, so it is an important concept. Each of these three ways of constructing the action has an experimental counterpart related to a fundamental piece or "bit" of quantum information called the *qubit*. This is important for understanding the information-theoretic reconstruction of quantum mechanics, since it builds the Hilbert space formalism of quantum mechanics from the qubit.

A qubit is fundamental in terms of information because it represents a question with a binary answer. For example, in our quantum gloves experiment the question is, "What is the spin of the glove in this direction of the dial?" and the answer is +1 or −1. So, the spin of a quantum glove is one example of a qubit. The real property of spin in quantum mechanics that we will introduce in Chapter 2 (spin-$\frac{1}{2}$) is a form of angular momentum and it has outcomes of $+h/4\pi = +\hbar/2$ or $-h/4\pi = -\hbar/2$ as the binary answers to, "What is the spin of this particle in the direction of the magnetic field?"

If the particle is deflected towards the North pole of the magnetic field the particle is said to have spin "up," and if it is deflected towards the South pole of the magnetic field it is said to have spin "down." This is the Stern–Gerlach experiment. The qubit formalism of quantum mechanics in this case gives us the up–down distribution of these pointlike events in space corresponding to the direction of the magnetic field for any given quantum state.

Photons also have spin angular momentum ($+\hbar$ or $-\hbar$) associated with their (circular) polarization (Chapter 7). Recall the concept of linear polarization of electromagnetic radiation from basic physics whereby the intensity of the waves is reduced according to the alignment of the polarizer with respect to the polarization of the waves. In the quantum case, a quantum of electromagnetic radiation (photon) either passes through the polarizer or does not. Therefore, the energy E it carries (as given by $E = hf$ where f is its frequency) is either passed or is not. So, the qubit in this case is the binary answer "yes" or "no" to the question, "Did the photon pass (was its energy transmitted) through the polarizer aligned in this direction?" The qubit formalism gives us the distribution of these pointlike events in time corresponding to the alignment of the polarizer for any given quantum state.

Finally, in Subchapter 9C, we explain how the double-slit experiment is a physical instantiation of the qubit associated with momentum multiplied by position. In other words, as the detector screen is translated relative to the double slits, the interference pattern associated with a momentum measurement $p = h/\lambda$, where λ is the wavelength of the quantum 'passing through' the slits, turns into a particle pattern associated with a "which slit" (position) measurement. So, the qubit in this case is the binary answer "left" or "right" to the question, "Which slit did the quantum pass through in contributing to the pattern on the detector screen?" The qubit formalism gives us the distribution of these pointlike events in space along the detector screen corresponding to the spatial location of the screen relative to the lens and slits for any given quantum state.

Since velocity generates boosts, angular momentum generates rotations, momentum generates spatial translations, and energy generates temporal translations, and bodily objects have worldtubes in spacetime as given by special relativity (Minkowski spacetime per NPRF + c) and interact via the exchange of quanta as given by quantum mechanics (Hilbert space per NPRF + h), an ontology based on the Poincaré symmetries is very natural. And, the Poincaré symmetries are very important for the dynamics of 'simple' bodily objects (particles) in quantum field theory. We apologize for this terse overview, but we wanted to give you a heads up because we will be focusing on spin-$\frac{1}{2}$ all the way through Chapter 7 and we do not want you to think that spin-$\frac{1}{2}$ is the only important physical instantiation of the qubit.

1.6.2 Simple, Beautiful, and Compelling

As we stated earlier, some in the foundations community believe entanglement is telling us that quantum mechanics is incomplete and/or inconsistent with special relativity. For example, entanglement led Maudlin to write [27, p. 23]:

> We cannot simply accept the pronouncements of our best theories, no matter how strange, if those pronouncements contradict each other. The two foundation stones of modern physics, Relativity and quantum theory, appear to be telling us quite different things about the world.

But, our principle explanation of entanglement is telling us just the opposite. That is, quantum mechanics is as complete as special relativity and, far from being inconsistent with special relativity, quantum mechanics harbors the exact same relativity principle at its foundation. To appreciate that consistency, we simply have to abandon our constructive bias.

So, in response to Wheeler, "the necessity of the quantum in the construction of existence" arises from the "simple, beautiful, and compelling" principle of NPRF. The lesson learned, i.e., the correction to the scientific worldview needed to resolve the EPR and EPR–Bell paradoxes, is that we must prioritize principle explanation over constructive explanation at the foundation of physics. In short, Einstein could have used the relativity principle to resolve (or even avoid) his EPR paradox exactly as he used it to solve the mystery of length contraction. Ironically, when it came to quantum mechanics, he chose Reichenbach's Principle over his beloved relativity principle.

The take-home message is that we are not introducing any new physical entities here, nor will it be necessary to modify or add anything to quantum mechanics. What is needed is a new principle way to conceive of quantum mechanics that was 'in principle' already at Einstein's disposal. We are simply proposing a new way to understand the successful physics we already have. And, according to this view, while physics needs to be completed (we need a theory of quantum gravity), the physics we have, if properly understood, is very comprehensive and coherent [38].

1.7 Summary

To summarize, quantum information theorists have reconstructed quantum mechanics in terms of information-theoretic principles. The empirically discovered fact at the foundation of their principle theory is (in one form or another) Information Invariance & Continuity. What we will show in this book is that Information Invariance & Continuity entails the invariant measurement of Planck's constant h between inertial reference frames related by spatial rotations or translations (Planck postulate), making their empirically discovered fact analogous to the light postulate of special relativity. Just as physicists tried unsuccessfully to account for the light postulate constructively via the luminiferous aether, physicists have tried unsuccessfully to account for the Planck postulate constructively by violating locality, statistical independence, intersubjective agreement, and/or unique experimental outcomes.

Consequently, we will follow Einstein's (very successful) lead in creating a principle *explanation* by justifying the light postulate with the relativity principle. That is, having shown that Information Invariance & Continuity entails the Planck postulate we will propose an obvious completion of the information-theoretic reconstruction program (a principle theory) via principle explanation by justifying Information Invariance & Continuity (Planck postulate) with the relativity principle. With this principle explanation we understand that quantum mechanics is based on the relativity principle (Hilbert space per NPRF + h) as opposed to dynamical laws and/or mechanistic causal processes, just like special relativity (Minkowski spacetime per NPRF + c).

But, wait a minute. Quantum mechanics is not Lorentz invariant, i.e., it is non-relativistic, so how can it be based on the relativity principle? As we explained above, the relativity principle is not restricted to "The laws of classical physics," so it can be applied to all of physics. Consequently, that it resides at the foundation of a theory does not mean the theory is 'relativistic'. For example, we showed how NPRF is at the foundation of Newtonian mechanics with its Galilean transformations, yet Newtonian mechanics is certainly non-relativistic.

The bottom line is, by prioritizing principle explanation over constructive explanation, e.g., prioritizing the relativity principle over Reichenbach's Principle, quantum mechanics can be as clearly understood as special relativity. So now we can answer Mermin's question [29], "What the hell are we talking about when we use quantum mechanics?":

> We are talking about the interacting bodily objects of our empirical investigations based on the invariance of Planck's constant h between the inertial reference frames of our data collection devices in accord with the relativity principle.

Contrary to consensus opinion, Bell's theorem does not toll for locality, statistical independence, etc., it only need toll for constructive explanation.

In conclusion, whether or not you still believe that the EPR facts are "self-contradictory" and that the quantum-mechanical violation of Bell's inequality is "absurd" after reading our principle solution to the mystery of entanglement depends on how strongly you are committed to the fundamentality of constructive explanation. Regardless, special relativity is understood in principle fashion via NPRF + c and you do not hear Nobel Laureates in Physics saying "nobody understands special relativity." Since information-theoretic reconstructions of quantum mechanics now allow us to understand quantum mechanics in principle fashion via NPRF + *h*, perhaps Nobel Laureates in Physics will soon stop saying "nobody understands quantum mechanics." With that overview of the book, we begin our story by introducing the mystery of entanglement for the general reader.

References

[1] E. Adlam, *The Problem of Confirmation in the Everett Interpretation*, Studies in History and Philosophy of Science Part B: Studies in History and Philosophy of Modern Physics, 47 (2014), pp. 21–32.
[2] E. Adlam, *Does Science Need Intersubjectivity? The Problem of Confirmation in Orthodox Interpretations of Quantum Mechanics*, Synthese, 200 (2022), p. 522.
[3] E. Adlam, 2023. Personal correspondence.
[4] E. Adlam and C. Rovelli, *Information is Physical: Cross-Perspective Links in Relational Quantum Mechanics*, 2022. Preprint. https://arxiv.org/abs/2203.13342.
[5] J. Bell, *On the Einstein–Podolsky–Rosen paradox*, Physics, 1 (1964), pp. 195–200.
[6] J. Bell, *Speakable and Unspeakable in Quantum Mechanics*, Cambridge University Press, Cambridge, 2nd ed., 2004.
[7] M. Born, A. Einstein, and I. Born, *The Born Einstein Letters: Correspondence between Albert Einstein and Max and Hedwig Born from 1916 to 1955 with Commentaries by Max Born*, Macmillan Press, London, 1971. Translated by Irene Born.
[8] J. Bricmont, *What Did Bell Really Prove?*, in Quantum Nonlocality and Reality, M. Bell and S. Gao, eds., Cambridge University Press, Cambridge, 2016, pp. 49–78.
[9] C. Brukner and A. Zeilinger, *Information Invariance and Quantum Probabilities*, Foundations of Physics, 39 (2009), pp. 677–689.
[10] O. Darrigol, *Relativity Principles and Theories from Galileo to Einstein*, Oxford University Press, New York, 2022.
[11] O. Darrigol, Personal Correspondence 2023.
[12] A. Einstein, *What is the Theory of Relativity?*, London Times, 28 November (1919), pp. 53–54.
[13] A. Einstein, *Physics and Reality*, Journal of the Franklin Institute, 221 (1936), pp. 349–382.
[14] A. Einstein, *Autobiographical notes*, in Albert Einstein: Philosopher-Scientist, P. Schilpp, ed., Open Court, La Salle, IL, 1949, pp. 3–94.
[15] A. Einstein, *Reply to Criticisms*, in Albert Einstein: Philosopher-Scientist, P. Schilpp, ed., Harper and Row, New York, 1959, p. 681.
[16] A. Einstein, B. Podolsky, and N. Rosen, *Can quantum-mechanical description of physical reality be considered complete?*, Physical Review, 47 (1935), pp. 777–780.
[17] L. Felline, *Scientific explanation between principle and constructive theories*, Philosophy of Science, 78 (2011), pp. 989–1000.
[18] L. Felline, *Mechanisms Meet Structural Explanation*, Synthese, 195 (2018), pp. 99–114.

[19] L. Felline, *Quantum theory is not only about information*, Studies in History and Philosophy of Science Part B: Studies in History and Philosophy of Modern Physics, (2018), pp. 1355–2198.

[20] L. Felline, *On Explaining Quantum Correlations: Causal vs. non-causal*, Entropy, 23 (2021), p. 589.

[21] R. Feynman, *Probability and Uncertainty – The Quantum Mechanical View of Nature*, 1964. https://www.youtube.com/watch?v=41Jc75tQcB0.

[22] G. Galilei, *Dialogue Concerning the Two Chief World Systems; Second Day*, University of California Press, Oakland, CA, 1953 [1632]. Translated by S. Drake.

[23] J. Hance and S. Hossenfelder, *Bell's theorem allows local theories of quantum mechanics*, Nature Physics, 18 (2022), p. 1382.

[24] P. Höhn, *Reflections on the information paradigm in quantum and gravitational physics*, Journal of Physics: Conference Series, 880 (2017), p. 012014.

[25] S. Hossenfelder, *Does Superdeterminism save Quantum Mechanics?*, 2021. https://www.youtube.com/watch?v=ytyjgIyegDI.

[26] D. Maltrana, M. Herrera, and F. Benitez, *Einstein's Theory of Theories and Mechanicism*, International Studies in the Philosophy of Science, 35 (2022), pp. 153–170.

[27] T. Maudlin, *Quantum Non-Locality and Relativity*, Wiley-Blackwell, Oxford, 2011.

[28] T. Maudlin, *What Bell Did*, Journal of Physics A, 47 (2014), p. 424010.

[29] N. D. Mermin, *Making better sense of quantum mechanics*, Reports on Progress in Physics, 82 (2019), p. 012002.

[30] P. Moylan, *Velocity reciprocity and the relativity principle*, American Journal of Physics, 90 (2022), pp. 126–134.

[31] J. Norton, *Einstein's Special Theory of Relativity and the Problems in the Electrodynamics of Moving Bodies That Led Him to It*, in The Cambridge Companion to Einstein, M. Janssen and C. Lehner, eds., Cambridge University Press, Cambridge, 2014, pp. 72–102.

[32] NOVA Season 46 Episode 2, *Einstein's Quantum Riddle*, 2019. https://www.pbs.org/video/einsteins-quantum-riddle-ykvwhm/.

[33] NOVA The Fabric of The Cosmos, *The Illusion of Distance and Free Particles: Quantum Entanglement*, 2013. https://www.youtube.com/watch?v=ZNedBrG9E90.

[34] PBS Space Time, *The Speed of Light is NOT About Light*, 2015. https://www.youtube.com/watch?v=msVuCEs8Ydo.

[35] C. Rovelli, *Relational quantum mechanics*, International Journal of Theoretical Physics, 35 (1996), pp. 1637–1678.

[36] M. Silberstein, W. M. Stuckey, and T. McDevitt, *Beyond Causal Explanation: Einstein's Principle Not Reichenbach's*, Entropy, 23 (2021), p. 114.

[37] L. Smolin, *Einstein's Unfinished Revolution: The Search for What Lies Beyond the Quantum*, Penguin Press, New York, 2019.

[38] W. M. Stuckey, T. McDevitt, and M. Silberstein, *"Mysteries" of Modern Physics and the Fundamental Constants c, h, and G*, Quanta, 11 (2022), pp. 5–14.

[39] W. M. Stuckey, T. McDevitt, and M. Silberstein, *No Preferred Reference Frame at the Foundation of Quantum Mechanics*, Entropy, 24 (2022), p. 12.

[40] W. M. Stuckey, M. Silberstein, and T. McDevitt, *Completing the Quantum Reconstruction Program via the Relativity Principle*, 2024. http://arxiv.org/abs/2404.13064.

[41] W. M. Stuckey, M. Silberstein, T. McDevitt, and I. Kohler, *Why the Tsirelson Bound? Bub's Question and Fuchs' Desideratum*, Entropy, 21 (2019), p. 692.

[42] W. M. Stuckey, M. Silberstein, T. McDevitt, and T. D. Le, *Answering Mermin's challenge with conservation per no preferred reference frame*, Scientific Reports, 10 (2020), p. 15771.

[43] L. Vaidman, *Why the Many-Worlds Interpretation?*, 2022. Preprint. https://arxiv.org/abs/2208.04618.

[44] J. A. Wheeler, *How Come the Quantum?*, New Techniques and Ideas in Quantum Measurement Theory, 480 (1986), pp. 304–316.

[45] A. Whitaker, *John Bell and the Most Profound Discovery of Science*, 1998. https://physicsworld.com/a/john-bell-profound-discovery-science/.

[46] L. Wolpert, *The Unnatural Nature of Science*, Harvard University Press, Cambridge, MA, 1993.

[47] A. Zeilinger, *A Foundational Principle for Quantum Mechanics*, Foundations of Physics, 29 (1999), pp. 631–643.

2

The Mystery: Mermin's Device

> One of the most beautiful papers in physics that I know of is yours in
> the *American Journal of Physics* **49** (1981) 10.
>
> Richard Feynman to David Mermin (30 March 1984)

So far, we have explained the mystery of entanglement by 'average-only' conservation in a fictitious quantum gloves experiment. While this works well to
convey the mystery of entanglement based on superposition to the general
reader, it lacks a derivation of the Bell inequality for the general reader that is
needed to fully appreciate the EPR–Bell paradox. In this chapter, we will share
Mermin's famous introduction to the mystery of entanglement referenced in
the Feynman quote above [5, p. 366–367]. In Mermin's presentation, Bell's
inequality and the quantum-mechanical violation thereof are clearly explained
for the general reader. Even if you are very familiar with the mystery of entanglement, we ask that you read this chapter since the rest of the book refers
extensively to this particular simplified instantiation of the mystery.

In two 1981 papers [11, 12], Mermin introduces a device (Figure 2.1) and
explains how it is operated and what it produces. According to Mermin, his
device [12] "should be regarded as something between a parable and a lecture
demonstration" because he did not actually build the device, but it is something that could be built if quantum mechanics is true. Herein, the Mermin
device represents a theoretical prediction of quantum mechanics concerning
a pair of entangled particles. Since quantum mechanics has been tested extensively and used successfully in all sorts of technologies, physicists generally
assume Nature behaves according to all of its predictions, to include those for
entangled particle pairs.

Indeed, as we pointed out in the Preface, the 2022 Nobel Prize in Physics
was awarded "for experiments with entangled photons, establishing the violation of Bell inequalities and pioneering quantum information science." In
other words, the mystery of entanglement is experimentally well established,
so the overwhelming majority of physicists accept the predictions of quantum

Einstein's Entanglement. W. M. Stuckey, Michael Silberstein, and Timothy McDevitt, Oxford University Press.
© Oxford University Press (2024). DOI: 10.1093/9780198919698.003.0003

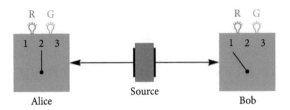

Fig. 2.1 The Mermin device. Alice has her detector on the left set to 2 and Bob has his detector on the right set to 1. The particles have been emitted by the source in the middle and are en route to the detectors.

mechanics as portrayed by the Mermin device. With that caveat out of the way, let's look at the device.

2.1 The Mysterious Facts of the Device

You don't need to understand anything about physics to appreciate the mystery of Mermin's device. It contains a source (the middle box in Figure 2.1) that emits a pair of particles towards two detectors (the boxes on the left and right in Figure 2.1) in each trial of the experiment. The settings (1, 2, or 3) on the left and right detectors are controlled randomly by Alice and Bob, respectively, i.e., we are assuming this experiment satisfies statistical independence. Each measurement at each detector produces either a result of R (red) or G (green) in each trial of the experiment. Here are the two facts that produce the mystery:

Fact 1 When Alice and Bob's settings happen to be the same in a given trial ("case (a)"), their outcomes are always the same: $\frac{1}{2}$ of the time RR (Alice's outcome is R and Bob's outcome is R) and $\frac{1}{2}$ of the time GG (Alice's outcome is G and Bob's outcome is G).

Fact 2 When Alice and Bob's settings happen to be different in a given trial ("case (b)"), the outcomes are the same $\frac{1}{4}$ of the time: $\frac{1}{8}$ RR and $\frac{1}{8}$ GG.

If you have read other popular accounts of entanglement, you probably recognize Fact 1 or its equivalent (as in Chapter 0). Fact 1 is sometimes presented as the mystery of entanglement, i.e., since the two detectors can be as far apart as we like, how is it that the measurement of one particle instantly establishes the outcome for the same measurement on the other particle? But, as Mermin points out, it is easy to explain Fact 1 causally without any "spooky actions at a distance." It is simply the case that the particles left the source with the same properties for each detector setting, what Mermin calls *instruction sets*. Fact 1 and the need for instruction sets characterize the EPR paradox.

For example, if Alice's particle has the instruction set RRG, her particle will produce an R outcome if she measures it in setting 1, an R outcome if she measures it in setting 2, and a G outcome if she measures it in setting 3. These are the hidden variables Einstein invoked in order to avoid "spooky actions at a distance." So, to guarantee that Bob's measurement outcomes will always agree with Alice's outcomes when they happen to choose the same detector setting (Fact 1), all we need is for Bob's particle to have the same instruction set as Alice's particle in every trial, e.g., when Alice's particle has the instruction set RRG, Bob's particle also has the instruction set RRG. In other words, instruction sets allow us to keep locality in accounting for Fact 1. Mermin emphasizes the locality assumption when he writes [11]:

> Why do the detectors always flash the same colors when the switches are in the same positions? Since the two detectors are unconnected there is no way for one to 'know' that the switch on the other is set in the same position as its own.

By "unconnected" Mermin means the detectors share no information about their settings and outcomes, superluminally or otherwise. Mermin writes [11]:

> This hypothesis, that the particles in a run carry identical instruction sets, is an obvious way to account for what happens in case (a). It cannot be proved that there is no other way, but I challenge the reader to suggest any.

What he means is that without hidden variables, one is left with nonlocality to explain Fact 1 (again, assuming statistical independence).

Mermin's instruction sets satisfy what is called *local realism*. That is, the outcome at Bob's measurement device is determined independently of what happens at Alice's measurement device (locality) and the measurement device is simply reacting to a property of the particle that exists independently of whether or not Alice or Bob chooses to measure it (realism). Mermin points out that the instruction sets can represent some properties of the particle and the measurement device in accord with some constructive explanation of why the particle–detector interaction produces the R and G measurement outcomes (realism). And, the causal mechanisms involved in that explanation do not require information exchanged between spacelike separated events (locality).

You will also see the term *separability* used instead of realism, i.e., someone might say that Mermin's instruction sets represent locality and separability.

That is, Einstein assumed the entangled particles are spatially separated, as he writes to Born in this March 1948 letter [2]:

> That which really exists in B should ... not depend on what kind of measurement is carried out in part of space A; it should also be independent of whether or not any measurement at all is carried out in space A. If one adheres to this program, one can hardly consider the quantum-theoretical description as a complete representation of the physically real. If one tries to do so in spite of this, one has to assume that the physically real in B suffers a sudden change as a result of a measurement in A. My instinct for physics bristles at this.

The spatially separated particles possess certain properties that dictate (in conjunction with properties of the detectors) the measurement outcomes on them. And, those facts about the particles hold true whether or not any measurement is actually performed on them. Abraham Pais wrote [13]:

> We often discussed his notions on objective reality. I recall that during one walk Einstein suddenly stopped, turned to me and asked whether I really believed that the moon exists only when I look at it.

From his sarcastic question, we can infer that Einstein would want to view the quantum gloves on a par with the classical gloves in our fictitious experiments.

So, for example, if Alice measures her particle in setting 2 and the outcome is R, then we know Bob's particle has the R property for setting 2 in that trial, regardless of whether or not Bob actually measures his particle in setting 2 for that trial. Again, this is exactly what Einstein desired of physics when he wrote, "physics should represent a reality in time and space." The second-to-last sentence in Einstein's 1948 quote above means Alice's measurement outcome of R cannot affect anything in Bob's measurement in a faster-than-light fashion, i.e., if physics is not "free from spooky actions at a distance," then Einstein's "instinct for physics bristles."

2.2 Bell's Inequality

But, as Bell showed nine years after Einstein died, quantum mechanics is not always consistent with this notion of local realism. Bell proved that precisely as Mermin does, i.e., by extending the EPR argument from case (a) to include the implications of instruction sets for case (b). Let's look at Mermin's reasoning.

Consider all trials for which the particles have the instruction set RRG, for example. Then, assuming Alice and Bob are randomly choosing their detector settings, Alice will chose setting 1 and Bob will choose setting 1 (detector setting pair "11") as often as Alice chooses setting 1 and Bob chooses setting 2 (detector setting pair "12"), ..., for this instruction set (and any other). In other words, any of the nine detector setting pairs (11, 12, 13, 21, 22, 23, 31, 32, 33) will be chosen as often as any other for this instruction set (and any other).[1] As a consequence, Alice and Bob will obtain the same outcomes in $\frac{1}{3}$ of those trials when they happen to choose different settings (case (b)) for the instruction set RRG. That is, there are six case (b) setting pairs (12, 13, 21, 23, 31, 32), and two of them (12, 21) will produce the same outcome (RR) while the other four (13, 23, 31, 32) will produce different outcomes (RG or GR).

Notice that this $\frac{1}{3}$ agreement for case (b) trials holds for any instruction set with two R (G) and one G (R). The only other type of instruction set is RRR or GGG, and in those trials the case (b) agreement is 100%. Therefore, the instruction sets necessary to explain Fact 1 per locality violate Fact 2. Specifically, Fact 2 for the Mermin device says we only get agreement in $\frac{1}{4}$ of case (b) trials, but instruction sets produce agreement in at least $\frac{1}{3}$ of case (b) trials. Indeed, if the instruction sets are produced with equal frequency, they will produce agreement in $\frac{1}{2}$ of case (b) trials, i.e., the outcomes are totally uncorrelated for case (b) trials just as in the classical gloves experiment in Chapter 0.

That instruction sets produce agreement in at least $\frac{1}{3}$ of case (b) trails is our *Bell inequality* and it is violated by Mermin's device. Since the Mermin device represents quantum mechanics and instruction sets represent local realism, we see that the quantum-mechanical outcomes must violate one or more of Bell's constructive assumptions, i.e., hidden variables, locality, or statistical independence, as we explained in Chapter 1.

So, the mystery of entanglement per the Mermin device can be summed up by this question: What is the constructive or principle explanation for Facts 1 and 2 of the Mermin device? Constructive attempts to answer that question (those that assume hidden variables) include faster-than-light causation (nonlocality, or "spooky actions at a distance"), backward-in-time causation (retrocausality), conspiratorial causal links between detector settings and the hidden variables (superdeterminism), no intersubjective agreement, and no unique experimental outcomes, as we have stated. We will discuss those in the

---

[1] Again, this is called statistical independence and we will explore the violations of this assumption for the Mermin device in the next chapter.

next chapter and explain why none has received strong support in the foundations community. In order to explain Facts 1 and 2 in principle fashion via NPRF using Information Invariance & Continuity from quantum information theory (Chapters 6 and 7), we will relate the Mermin device to the measurement of spin for spin-$\frac{1}{2}$ particles (like the electron).

As we stated in Chapter 1, the information-theoretic reconstructions of quantum mechanics build the (finite-dimensional) Hilbert space formalism of quantum mechanics from the binary answer, e.g., yes–no, up–down, etc., to some question, just as in classical probability theory. In quantum probability theory, this fundamental unit of information is called a quantum bit, or *qubit* for short. The strength of information theory lies in its generality, i.e., facts are established regardless of how the information might be instantiated physically. Indeed, Höhn writes [9]:

> ... one could argue that quantum reconstructions come a substantially longer way and from more elementary assumptions than the derivation of special relativity from the relativity principle,

although he admits that [10], "Entanglement from complementarity is not as intuitive as the relativity of simultaneity from the relativity principle." [Note: This actually misses the analogy, as we will explain in Chapter 6.]

That is, its generality is also its weakness if you do not already know what the information-theoretic terminology means. So, we will restrict our explanations to actual physical examples, which means our presentation is not as general as the actual reconstructions themselves, especially if you view information as fundamental. However, without relating the qubit to actual experiments, we cannot see how the relevant Poincaré symmetries result from NPRF + h. As we pointed out in Subsection 1.6.1, we will ultimately relate the qubit to three different experiments, but the primary example of a qubit we will use is the spin-$\frac{1}{2}$ particle, since, as Brukner and Zeilinger said [3], "spin-$\frac{1}{2}$ affords a model of the quantum mechanics of all two-state systems, i.e., qubits."

2.3 Spin Angular Momentum

The spin we are talking about here is a real particle property corresponding to the binary outcomes of our fictitious experiments (R or G, Black or White, Left or Right). Spin is generally characterized as the 'intrinsic angular momentum' of electrons (and some other particles). It was discovered accidentally in 1922 by Otto Stern and Walther Gerlach as they attempted to measure the orbital angular momentum of the silver atom due to its valence electron.

Earth is often used as an analogy to differentiate spin and orbital angular momentum, i.e., Earth spins on its axis once a day (spin angular momentum) and it orbits the Sun once a year (orbital angular momentum). Since electrons are charged particles, they should produce a magnetic field like that of a magnet when they rotate in any fashion, i.e., when they possess angular momentum. Therefore, rotating electrons should experience a magnet-to-magnet force when placed in an external magnetic field. You can feel the magnet-to-magnet force between a pair of everyday magnets when holding them in your hands and bringing them near each other. When the same poles are facing each other the magnets repel each other and when the opposite poles are facing each other the magnets attract each other.

Stern and Gerlach exploited this fact by creating a magnetic field that was stronger near one pole than the other and then passing a beam of silver atoms through the inhomogeneous magnetic field [7, 8]. Since the magnetic field of the Stern–Gerlach (SG) magnets is stronger near the North pole than the South pole, the North pole effect on the 'atomic magnet' (magnetic moment) dominates that of the South pole. Essentially, Stern and Gerlach imagined they were creating a beam of randomly oriented atomic magnets and passing that beam through an inhomogenous magnetic field to see how the individual silver atoms would be deflected. Their experiment led to three different predicted outcomes (Figures 2.2, 2.3, and 2.4).

Given the constructive model per classical physics of randomly oriented atomic magnets in an external magnetic field, Joseph Larmor expected to see the silver atoms deflected throughout the detector screen [7] (Figure 2.2). In

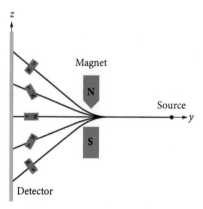

Fig. 2.2 The classical model of the Stern–Gerlach experiment. If the atoms enter with random orientations of their 'intrinsic' magnetic moments (due to their 'intrinsic' angular momenta), the SG magnets should produce all possible deflections.

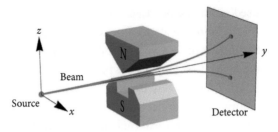

Fig. 2.3 An SG spin measurement showing the two experimental outcomes, up ($+\hbar/2$) and down ($-\hbar/2$) relative to the poles of the SG magnets, or $+1$ and -1 for short.

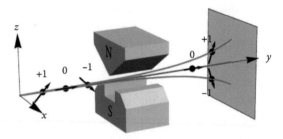

Fig. 2.4 The quantum prediction for the Stern–Gerlach experiment per Bohr. If the valence electron of the silver atom is in a circular orbit, Bohr expected its quantized orbital angular momentum to be projected parallel, opposite, or normal to the SG magnetic field. That means the beam of silver atoms would split into three parts.

contrast, the model of quantum theory that Stern and Gerlach believed they were testing was that of Arnold Sommerfeld, which predicted just two deflection directions [7] (up and down, Figure 2.3). As it turns out, that quantum expectation differed from what Neils Bohr believed should happen according to his Bohr model of the atom [8].

Bohr's model said the outermost (valence) electron would have a circular orbit, and Bohr thought the quantized orbital angular momentum of that orbit could be projected parallel, opposite, or normal to the magnetic field. Stern and Sommerfeld hypothesized that the quantized orbital angular momentum of that orbit would only be projected parallel or opposite to the magnetic field. So, Bohr's version of the Bohr model predicted the beam of silver atoms would be split three ways (Figure 2.4), while Stern and Sommerfeld's version of the Bohr model predicted the beam of silver atoms would be split only two ways (Figure 2.3). To Bohr's surprise, the Stern–Gerlach experiment produced just two deflections of the beam and he wrote to Gerlach [8]:

I would be very grateful if you or Stern could let me know, in a few lines, whether you interpret your experimental results in this way that the atoms are oriented only parallel or opposed, but not normal to the field, as one could provide theoretical reasons for the latter assertion.

Although Stern and Sommerfeld's prediction was correct, it was correct for the wrong reasons. The ultimate explanation of the Stern–Gerlach experiment was that the silver atoms were in the zero orbital angular momentum state, so the quantum result was not due to the valence electron's orbital angular momentum but to its spin angular momentum, which can take on two different values, $\pm h/4\pi$ or $\pm \hbar/2$, where h is Planck's constant and $\hbar = h/2\pi$.

The property of spin was not proposed until 1925 by George Uhlenbeck and Samuel Goudsmit [14], and even then they did not relate it to the Stern–Gerlach experiment. Bretislav Friedrich and Dudley Herschbach write [8]:

A curious historical puzzle remains. In view of the interest aroused by the [Stern–Gerlach experiment] in 1922, we would expect that the postulation of electron spin in 1925 should very soon have led to a reinterpretation of the [Stern–Gerlach experiment] splitting as really due to spin. However, the earliest attribution of the splitting to spin that we have found did not appear until 1927, when Ronald Fraser noted that the ground-state orbital angular momentum and associated magnetic moments of silver, hydrogen, and sodium are zero. Practically all current textbooks describe the Stern–Gerlach splitting as demonstrating electron spin, without pointing out that the intrepid experimenters had no idea it was spin that they had discovered.

2.4 The Planck–Einstein Relation

Planck introduced his constant h in 1900 to solve a mystery that existed in the late 1800s, i.e., blackbody radiation. You are probably familiar with the fact that heated objects will glow red, thus the term *red hot*. As the temperature increases, the color of emitted light from the object turns white, thus the term *white hot*. That is, more high-frequency light is emitted as the temperature increases. Physicists measured the spectrum of light from heated objects (blackbody radiation), but they didn't have an equation fitting that spectrum as a function of temperature. Just as the classical model for spin (Figure 2.2) does not match the Stern–Gerlach experiment (Figure 2.3), the classical model for blackbody radiation did not match the measured blackbody spectrum.

According to the classical model for blackbody radiation, adding heat to an object made its charged constituents vibrate more vigorously (kinetic theory of matter). Maxwell's equations for light then predicted that those vibrating charged particles should emit light (or, more generally, electromagnetic radiation). But when physicists used this classical model, their equations predicted much more high-frequency radiation at a given temperature than was observed.

Planck solved this problem heuristically by assuming something that Maxwell's equations did not predict, i.e., higher-frequency light contained more energy than lower-frequency light. He did that by assuming the oscillating constituents of the blackbody could only emit or absorb light in discrete amounts (which he called quanta) with energy E proportional to the frequency f of the oscillators, $E = hf$. Accordingly, Einstein then assumed light must exist in quanta (called photons) with $E = hf$ and used that to explain the photoelectric effect in 1905 (which won him the 1921 Nobel Prize in Physics). The equation $E = hf$ is called the Planck–Einstein relation and the constant of proportionality h is Planck's constant. [Planck actually determined the value of h by fitting Wien's law to experimental measurements before he derived his radiation law [4, 6].]

2.5 The Quantum-Mechanical Reconciliation

So, we have the following facts determined experimentally:

(i) Planck (1900) and Einstein (1905) discovered that electromagnetic radiation was emitted and absorbed in discrete amounts of energy $E = hf$, so that h is a constant in both of Planck's blackbody radiation and Einstein's photoelectric effect laws of Nature.

(ii) Stern and Gerlach (1922) discovered that a beam of silver atoms splits up and down relative to a magnetic field at all spatial orientations of the magnets. This quantum property of the atom's valence electron is called *spin angular momentum* and its value is given by Planck's constant, i.e., $\pm\hbar/2$.

Therefore, as Steven Weinberg pointed out, measuring an electron's spin via SG magnets constitutes the measurement of "a universal constant of nature, Planck's constant h" [15] per item (i). Recognizing that complementary spin measurements establish a reference frame (Chapter 6) and inertial reference frames are related by spatial rotations, item (ii) is therefore justified by NPRF.

As we will see in Chapter 6, this leads to the 'average-only' projection of spin angular momentum. That is, since fractional values of h are not allowed, a qubit of spin angular momentum cannot be further reduced by projection along a magnetic field as expected per the classically continuous projection of angular momentum.

In Chapter 7, we show how the invariance of h between inertial reference frames related by spatial rotations also leads to the 'average-only' transmission of linearly polarized photons through a polarizing filter. That is, since fractional values of h are not allowed, a qubit of polarized electromagnetic energy $E = hf$ cannot be further divided using a polarizing filter as expected per the classically continuous filtration of a polarized electromagnetic field.

Finally in Subchapter 9C, we show how the invariance of h between inertial reference frames related by spatial translations leads to wave–particle duality in the double-slit experiment. That is, since fractional values of h are not allowed, a qubit of momentum $p = h/\lambda$ cannot be spatially distributed along the detector of the double-slit experiment as expected per the classically continuous interference of waves.

In all three examples, NPRF + h demands that a classically continuous quantity (angular momentum, energy, momentum) be quantized, so that the classically continuous predictions follow from the quantum results on average. Since quantum information theorists derived the (finite-dimensional) Hilbert space of quantum mechanics from the qubit in general, we see that their principle account of quantum mechanics follows the same historical pattern as special relativity. Just as Maxwell's equations established a constant of Nature c, the Planck radiation law established a constant of Nature h. Both of these constants were subsequently found to have the same values in all inertial reference frames in accord with NPRF (the Michelson–Morley experiment for c and the Stern–Gerlach experiment for h), whence the principle theories of special relativity and quantum mechanics, respectively. NPRF then provides for a principle explanation of the predictions of these principle theories.

We will return to this point in Chapter 6, but let's continue mapping Mermin's device to a real physics experiment involving spin angular momentum. Spins in orthogonal directions are complementary variables like quantum color and quantum handedness in our imaginary quantum gloves experiment, so spin measurements can create an EPR–Bell paradox.

It turns out that processes exist in Nature whereby spin angular momentum is conserved in the emission of a pair of particles, e.g., the dissociation of a spin-zero diatomic molecule [1]. Generally speaking, the pair of particle spins in such processes can be anti-aligned (to sum to zero) or aligned (to sum to $\pm\hbar$). There are four quantum-mechanical states representing the pair of spin-entangled particles created in such processes.

The Bell spin singlet state (or *singlet state* for short) represents a maximally entangled pair of particles with anti-aligned spins in any direction of space. The three Bell spin triplet states represent a maximally entangled pair of particles with aligned spins in any direction of a particular spatial plane, e.g., the *xy*, *xz*, or *yz* planes with suitably chosen *x, y, z* axes. The plane in which a Bell spin triplet state (or *triplet state* for short) exhibits this behavior is called its *symmetry plane*. We will refer to these four states collectively as *Bell states*. [You do not need to worry about these mathematical nuances, you can follow the argument conceptually without the technical details.]

In the Mermin device, Alice and Bob obtain the same outcomes when they happen to choose the same detector settings, so the Mermin device maps most easily to a triplet state in its symmetry plane (hereafter, we will always assume the triplet state is being measured in its symmetry plane). The three choices of detector settings on the Mermin device correspond to three different SG magnet orientations in the symmetry plane, where Alice's SG magnet orientation is given by $\hat{a}$ and Bob's is given by $\hat{b}$ (Figures 2.5 and 2.6). In his paper, Mermin maps his device to the singlet state by having the meaning of R and G for Alice be opposite that for Bob (like we did for the quantum handedness outcomes in our quantum gloves experiment). We will just stick to the triplet state where the mapping between R–G and spin up–spin down is direct.

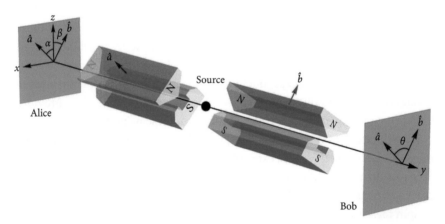

Fig. 2.5 Alice and Bob making spin measurements on a pair of spin-entangled particles with their SG magnets and detectors.

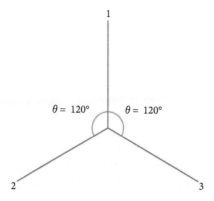

Fig. 2.6 Three possible orientations of Alice and Bob's SG magnets for the Mermin device.

Quantum mechanics says the probability of Alice and Bob obtaining the same results as a function of the angle θ between $\hat{a}$ and $\hat{b}$ for the triplet state is $P(++) = P(--) = \frac{1}{2}\cos^2(\theta/2)$. And, quantum mechanics says the probability of Alice and Bob obtaining different results as a function of the angle θ between $\hat{a}$ and $\hat{b}$ for the triplet state is $P(+-) = P(-+) = \frac{1}{2}\sin^2(\theta/2)$. So, when $\theta = 0$ (case (a)), quantum mechanics says $P(++) = P(--) = \frac{1}{2}$ and $P(+-) = P(-+) = 0$. Letting R denote spin up (+) and G denote spin down (−), we obtain Fact 1 for the Mermin device. When $\theta = 120°$ (case (b)), quantum mechanics says $P(++) = P(--) = \frac{1}{8}$ and we have Fact 2 for the Mermin device.

2.6 Mermin's Challenge

So, we see that the mystery of entanglement is easily conveyed by Facts 1 and 2 for the Mermin device, which represents the prediction by quantum mechanics for the correlated outcomes of measurements on spin-entangled particles in the triplet state. The problem is, quantum mechanics does not specify a reason, constructive or principle, for those correlations other than conservation of spin angular momentum and, as we saw in Chapter 0 and will see again in Chapter 7, the conservation of spin angular momentum entailed by the quantum-mechanical probabilities ('average-only' conservation) does not solve the mystery of entanglement, it *is* the mystery, i.e., it is what needs to be explained.

In the last sentence of his paper, Mermin issues the following challenge [11]:

It is left as a challenging exercise to the physicist reader to translate the elementary quantum-mechanical reconciliation of cases (a) and (b) into terms meaningful to a general reader struggling with the dilemma raised by the device.

In Chapter 7 we will answer Mermin's challenge by deriving the joint probabilities for "the elementary quantum-mechanical reconciliation of cases (a) and (b)" given above according to Mermin's rules, i.e., without violating locality, statistical independence, intersubjective agreement, or unique experimental outcomes. Since Bell's theorem proves this cannot be done constructively (i.e., using hidden variables), we will do this in principle fashion.

In an April 1948 letter to Born, Einstein wrote [2]:

Those physicists who regard the descriptive methods of quantum mechanics as definitive in principle would ... drop the requirement for the independent existence of the physical reality present in different parts of space; they would be justified in pointing out that the quantum theory nowhere makes explicit use of this requirement. I admit this, but would point out: when I consider the physical phenomena known to me, and especially those which are being so successfully encompassed by quantum mechanics, I still cannot find any fact anywhere which would make it appear likely that requirement will have to be abandoned. I am therefore inclined to believe that the description of quantum mechanics ... has to be regarded as an incomplete and indirect description of reality ...

This is in direct reference to the EPR paradox:

- Quantum mechanics is complete.
- There are no "spooky actions at a distance" (locality).

Our principle explanation will show that quantum mechanics is as complete as possible (as complete as special relativity) and it will not invoke "spooky actions at a distance" to solve the mystery of entanglement. Ironically, the compelling fundamental principle we will use to resolve the EPR paradox and answer Mermin's challenge is none other than Einstein's own relativity principle, NPRF.

2.7 Summary

The Mermin device shows clearly how hidden variables (instruction sets), locality, and statistical independence are assumed in deriving a Bell inequality. We will review the consequences of violating Bell's constructive assumptions in the next chapter. What we will find is an historical trend towards increasing 'desperation'.

References

[1] D. Bohm, *Quantum Theory*, Prentice-Hall, Hoboken, NJ, 1952.

[2] M. Born, A. Einstein, and I. Born, *The Born Einstein Letters: Correspondence between Albert Einstein and Max and Hedwig Born from 1916 to 1955 with Commentaries by Max Born*, Macmillan Press, London, 1971. Translated by Irene Born.

[3] C. Brukner and A. Zeilinger, *Operationally Invariant Information in Quantum Measurements*, Physical Review Letters, 83 (1999), pp. 3354–3357.

[4] J. Diaz, *This Math Trick Revolutionized Physics*, 2024. https://www.youtube.com/watch?v=gXeAp_lyj9s.

[5] M. Feynman, *Perfectly Reasonable Deviations from the Beaten Track: The Letters of Richard P. Feynman*, Basic Books, New York, 2005.

[6] M. Fowler, *Planck's Route to the Black Body Radiation Formula and Quantization*, 2009. https://galileo.phys.virginia.edu/classes/252/PlanckStory.htm.

[7] A. Franklin and S. Perovic, *Experiment in Physics: Appendix 5: Right Experiment, Wrong Theory: The Stern–Gerlach experiment*, in The Stanford Encyclopedia of Philosophy, E. N. Zalta, ed., Stanford University, 2019.

[8] B. Friedrich and D. Herschbach, *Stern and Gerlach: How a Bad Cigar Helped Reorient Atomic Physics*, Physics Today, 56 (2003), p. 53.

[9] P. Höhn, *Reflections on the information paradigm in quantum and gravitational physics*, Journal of Physics: Conference Series, 880 (2017), p. 012014.

[10] P. Höhn, *Complementarity Identities from an Informational Reconstruction*, 2023. Conference: The Quantum Reconstruction Program and Beyond. https://www.youtube.com/watch?v=60ZQ9Fp2cBo.

[11] N. Mermin, *Bringing home the atomic world: Quantum mysteries for anybody*, American Journal of Physics, 49 (1981), pp. 940–943.

[12] N. Mermin, *Quantum mysteries for anyone*, Journal of Philosophy, 78 (1981), pp. 397–408.

[13] A. Pais, *Einstein and the quantum theory*, Reviews of Modern Physics, 51 (1979), pp. 863–914.

[14] G. Uhlenbeck and S. Goudsmit, *Spinning Electrons and the Structure of Spectra*, Nature, 117 (1926), pp. 264–265.

[15] S. Weinberg, *The Trouble with Quantum Mechanics*, The New York Review of Books, 19 January (2017). https://www.nybooks.com/articles/2017/01/19/trouble-with-quantum-mechanics/.

3

Constructive Responses to Bell's Theorem

> On 4 November 1964, a journal called *Physics* received a paper written
> by John Bell, a theoretician from CERN. The journal was short-lived,
> but the paper became famous, laying the foundations for the modern
> field of quantum-information science.
>
> CERN News (4 November 2014)

Our fictitious quantum gloves experiment in Chapter 0 introduced the roles
played by superposition and continuity in establishing the mystery of entanglement. That mystery was characterized by 'average-only' conservation of an
imaginary quantum glove property called spin. The Mermin device in the last
chapter introduced the quantum-mechanical violation of the Bell inequality,
which follows from the assumptions of locality and statistical independence
for a constructive approach, as well as the real quantum property of spin.
Thus, we have thoroughly introduced the concepts needed to understand the
EPR–Bell paradox arising from the mystery of entanglement.

In this chapter we explore some constructive efforts to solve the greatest
mystery in physics. We are not here to argue for or against these constructive
approaches, we leave that for the proponents and opponents themselves. We
are simply using some selected examples to convey how they violate locality,
statistical independence, intersubjective agreement, or unique experimental
outcomes, so the reader can appreciate the consequences of such violations.

The title of Bell's 1964 paper in *Physics* is "On the Einstein–Podolsky–Rosen
paradox" [10]. Again, his theorem/inequality published in that paper has been
called "the most profound discovery of science," and David Kaiser said [34]:

> We now know with hindsight [Bell's paper] was one of the most significant
> articles in the history of physics—not just the history of 20th century physics,
> in the history of the field as a whole.

So, we begin this chapter with a quick review of the EPR argument and
Bell's inequality derived from his constructive assumptions, which give rise

Einstein's Entanglement. W. M. Stuckey, Michael Silberstein, and Timothy McDevitt, Oxford University Press.
© Oxford University Press (2024). DOI: 10.1093/9780198919698.003.0004

to the EPR and EPR–Bell paradoxes, respectively. Then we will explain the constructive bias and give examples of the violations of locality, statistical independence, intersubjective agreement, and unique experimental outcomes that result from constructive attempts to resolve the EPR–Bell paradox. That should make it clear why constructive accounts of entanglement have not garnered consensus support in the foundations community.

What our (selectively abbreviated) history of such constructive accounts will show are increasingly 'desperate' models of reality required to understand quantum mechanics constructively. As we explained in Chapter 1, this is analogous to efforts to save geocentricism before Galileo invoked NPRF, and efforts to save the constructive explanation of length contraction before Einstein invoked NPRF.

3.1 Revisiting the EPR Paradox

In "Can Quantum-Mechanical Description of Physical Reality be Considered Complete?" Einstein, Podolsky, and Rosen make a very simple case that there exist "elements of reality" without counterparts in the formalism of quantum mechanics. Therefore, EPR concluded that quantum mechanics is either incomplete or Nature is nonlocal (again, they assumed statistical independence). Since "spooky actions at a distance" (nonlocality) were unacceptable to Einstein, EPR concluded that quantum mechanics is incomplete. Let's look again at that argument.

EPR dealt with the complementary variables of position and momentum for a pair of particles that were created so as to conserve momentum, i.e., the particles are momentum entangled. For example, suppose the particles have the same mass and move in opposite directions with the same speed away from their common source, then their momenta after emission sum to zero and that equals their momenta before emission.

According to EPR, if we measure the position of particle 1, then we know the exact position of particle 2 with 100% certainty (probability = 1) because particle 2 was emitted at the same time from the same place and is moving at the same speed in the opposite direction as particle 1, i.e., the particles are also position entangled. EPR said the position of particle 2 is an "element of reality" because we can know it exactly with probability = 1 without actually measuring it. That is, the position of particle 2 is a fact about reality regardless of whether or not we actually measure it. Again, this is sometimes called realism (as in the instruction sets of Chapter 2) or *counterfactual definiteness*, it is what Einstein was referring to when he asked Pais whether or not he "believed that the moon exists only when I look at it." The obvious answer is "no," the moon is orbiting the Earth and interacting with other bodily objects whether or not we are actually making measurements of it, i.e., the position of the moon is an "element of reality."

We use realism / counterfactual definiteness in our subjective spacetime models of reality for everyday reasoning. For example, if you park your car outside on a sunny summer day, your car is absorbing radiation from the Sun according to your spacetime model of reality regardless of whether anyone is looking at it, videotaping it, or otherwise making measurements of it. Consequently, you do not leave your cool drink in your parked car when you arrive at work because it will be objectionably hot when you retrieve it for lunch.

Returning to the EPR argument, it is also the case that the measurement of particle 1 does not create the position of particle 2 at spacelike separation (locality). Again, Einstein summed up his belief in realism and locality with his statement [13], "physics should represent a reality in time and space, free from spooky actions at a distance." Such nonlocal causation was just too "spooky" for Einstein, as can be seen in this excerpt from his letter to Cornel Lanczos in 1942 [20]:

> It seems hard to sneak a look at God's cards. But that he plays dice and uses 'telepathic' methods (as the present quantum theory requires of him) is something that I cannot believe for a moment.

Here, "'telepathic' methods" is another pejorative reference to nonlocal influences.

Continuing with this line of reasoning, if we rather choose to measure the momentum of particle 1, then we know the exact momentum of particle 2 with 100% certainty because it is equal and opposite to particle 1's momentum, i.e., momentum is also an "element of reality." Therefore, the position and momentum of particle 2 surely exist simultaneously as "elements of reality," but quantum mechanics does not allow us to simultaneously know the exact position and momentum of a particle with certainty because they are complementary variables. So, quantum mechanics is either incomplete (its formalism does not account for "elements of reality" that exist to avoid nonlocality) or it is nonlocal (requires faster-than-light influences to establish correlated outcomes). We can state this in the form of a paradox as defined in Chapter 1 using these as our paradoxical EPR facts:

- Quantum mechanics is complete.
- There are no "spooky actions at a distance" (locality).

These two facts are "absurd or self-contradictory" because if quantum mechanics is complete, then no hidden variables (instruction sets) can be used to account for Fact 1 of the Mermin device, since unmeasured variables are not definite in quantum mechanics. But without those instruction sets, Fact 1 requires nonlocality to account causally for the correlation of the entangled particles according to Reichenbach's Principle (again, assuming statistical independence). That is, our two EPR facts are "self-contradictory" according to our scientific worldview. To resolve the paradox we must give up at least one of those EPR facts or conclude that the scientific worldview needs to be corrected so that the EPR facts are not seen as contradictory. Enter John Bell, a particle physicist by day and a foundations physicist by night.

Bell realized that the key to the EPR paradox was the fact that their definition of "an element of reality" is at odds with complementary variables in quantum mechanics. So, Bell extended the EPR argument from Fact 1 for case (a) to the consequences of case (b), i.e., Fact 2, exactly as shown by Mermin. Let's look more closely at the quantum formalism concerning complementary variables.

3.2 Complementary Variables

As we said, position and momentum are two such complementary variables (more on that in Chapter 9), and so are spins in orthogonal directions. According to quantum mechanics, if you first measure the position then measure the momentum of a particle, you will get different answers than if you first measure momentum then measure position. Mathematically, position x times momentum p does not equal momentum p times position x. The difference $xp - px$ is called the *commutator* and is given by Planck's constant h, i.e., $xp - px = ih$ where $i = \sqrt{-1}$ (quantum mechanics uses complex numbers).

This is what we meant in Chapter 0 when we said h establishes a limit to how much simultaneous information one can obtain about a quantum particle. As Höhn points out [27], Planck's constant represents "a universal limit on how much simultaneous information is accessible to an observer" because "thanks to the existence of complementarity, implied by $\hbar$, an observer may not access all conceivable properties of the system at once."[1]

This is very different from classical mechanics where one uses the initial position and momentum of an object exactly and simultaneously to determine its later position and momentum exactly and simultaneously. In classical mechanics position and momentum commute, i.e., position times momentum

---

[1] We will say more about this in Chapter 6 when we review the fundamentals of quantum information theory.

equals momentum times position. This is just one of many places in quantum mechanics where the formalism of quantum mechanics reduces to the formalism of classical mechanics by letting $h \to 0$. So we see that $h = 0$ means we have no limit on how much simultaneous information one can obtain about a classical particle. Indeed, it is often said that quantum mechanics is more fundamental than classical mechanics because $h \neq 0$. [Recall, we have a similar situation with the speed of light c: it is often said that relativistic mechanics is more fundamental than Newtonian mechanics because $c \neq \infty$.]

This is also conveyed by the Heisenberg uncertainty principle:

$$\Delta x \Delta p \geq \frac{\hbar}{2},$$

where Δx is the uncertainty in position and Δp the uncertainty in momentum. That is, the more we know about the position of a particle (Δx gets smaller), the less we know about its momentum (Δp gets larger), and vice versa. Indeed, if we know the position (momentum) of the particle exactly, then we know absolutely nothing about its momentum (position). According to Werner Heisenberg [38, p. 117], "even in principle, we cannot know the present in all detail." Let's apply this to the measurement of spin using Mermin's measurement device in Chapter 2.

Suppose Alice uses three Mermin-like measurement devices to perform a sequence of three spin measurements (Figure 3.1) on a spin-$\frac{1}{2}$ particle, i.e., the source in this case is producing just one particle instead of two as in the Mermin device. Further, suppose Alice first measures her particle in detector setting 2 and gets R, then uses the second measurement device to measure her particle in detector setting 1 and gets G. If the particle obeyed classical mechanics and Alice uses the third detector to again measure her particle in detector setting 2, she will again get R, since she has already determined the particle is R when measured in setting 2. But, quantum mechanics says outcome R is not guaranteed for that third measurement because the information contained by the particle is now G for setting 1, not R for setting 2. You cannot know both the value of spin in setting 1 and the value of spin in setting 2 simultaneously for a quantum particle like you can for a classical particle.

To compute the distribution of R–G information for setting 2 in the third measurement, we start with the unit-length vector of information we have, i.e., the quantum state |G for setting 1⟩ in Hilbert space (Figure 3.2). That the quantum state is the unit-length vector |G for setting 1⟩ along the G for setting 1 outcome axis is in total analogy to $\hat{x}$ being the unit-length vector along the x axis in real space. This unit-length vector |G for setting 1⟩ means that the

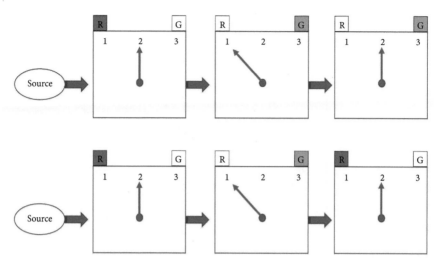

Fig. 3.1 Sequential spin measurements. Alice measures the spin of her particle in setting 2 and obtains R, then measures it in setting 1 and gets G. One might think that if she then measures it in setting 2, she will again get R, since the value of spin for setting 2 was already determined to be R in the first measurement. However, quantum mechanics says there is now a probability of 0.25 that she will obtain G for that third measurement, so the first configuration shown above will constitute $\frac{1}{4}$ of all such trials while the second configuration will constitute $\frac{3}{4}$ of all such trials.

outcome of a setting 1 measurement on this quantum state will produce G with probability $1^2 = 1$.

To find out how that information is distributed between R and G outcomes for a setting 2 measurement, quantum mechanics says you project the unit-length quantum state vector along the unit-length vectors on the outcome axes for setting 2, i.e., $|R$ for setting $2\rangle$ and $|G$ for setting $2\rangle$ (Figure 3.2), then square the results. Thus, the probability of a G outcome for the setting 2 measurement of the quantum state $|G$ for setting $1\rangle$ is

$$|\langle G \text{ for setting } 1 | G \text{ for setting } 2 \rangle|^2 = \left(\frac{1}{2}\right)^2 = \frac{1}{4},$$

and the probability of an R outcome for the setting 2 measurement of the quantum state $|G$ for setting $1\rangle$ is

$$|\langle G \text{ for setting } 1 | R \text{ forsetting } 2 \rangle|^2 = \left(\frac{\sqrt{3}}{2}\right)^2 = \frac{3}{4}.$$

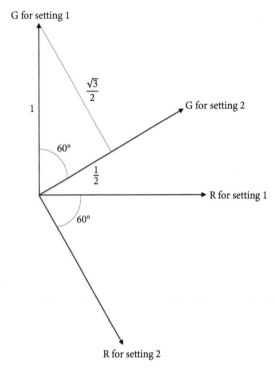

Fig. 3.2 Hilbert space for settings 1 and 2. Since the SG magnet orientations between settings 1 and 2 for a measurement of spin in the Mermin device differ by 120° (Figure 2.6), the angle between the setting 1 outcome axes and the setting 2 outcome axes in Hilbert space is 60°.

That is, the amount of information contained by a qubit for some binary-outcome measurement is fixed (invariant) and distributed according to the Pythagorean theorem between the binary outcomes of different measurements on that qubit. That is why there is a limit to the simultaneous knowledge we can possess about a qubit for different measurements on that qubit. So, spin measurements in different directions, e.g., S_x, S_y, and S_z, do not commute and, just like position and momentum, their commutator in orthogonal directions is given by Planck's constant, i.e., $S_x S_y - S_y S_x = i\hbar S_z$. As we will see in Chapter 6, this follows from NPRF + h or, in information-theoretic terms, the empirically discovered fact of Information Invariance & Continuity as justified by the relativity principle.

3.3 Bell's Assumptions

That is why the key to the EPR–Bell paradox per the Mermin device is not Fact 1, i.e., Alice and Bob always obtain the same outcomes (RR or GG) when they happen to choose the same detector settings (11, 22, 33). We can easily satisfy Fact 1 causally without violating locality or statistical independence by admitting that quantum mechanics is constructively incomplete and completing it using instruction sets. Rather, the EPR–Bell paradox stems from the fact that the instruction sets used to explain Fact 1 in accord with locality do not reproduce Fact 2, i.e., Alice and Bob obtain the same outcomes (RR or GG) in $\frac{1}{4}$ of all trials when they happen to choose different detector settings (12, 13, 21, 23, 31, 32). So, let's explore the assumptions behind Bell's theorem and see what kind of constructive account is required to explain Facts 1 and 2 of the Mermin device.

First, again, Bell and Mermin want a constructive account. That is, they assume the existence of hidden variables (instruction sets for the Mermin device) in order to explain the correlations of Fact 1 causally (as demanded by Reichenbach's Principle) in accord with locality (and statistical independence). Hidden variables are the cause of measurement outcomes and a complete (constructive) theory of quantum physics would provide the dynamical details for that causation, i.e., a constructive account of the Planck postulate à la an aether account of the light postulate. So, hidden variables stand for more than the simple exchange of energy–momentum between the source and detector. Clearly there is an exchange of energy–momentum between source and detector or the detector would not flash R or G at all.

For example, the complete theory might require information about properties of both the measurement device and the particle, and then invoke some dynamical law or mechanistic causal process that determines the measurement outcome from those properties. The instruction sets represent all of that behind-the-scenes machinery. Again, there is an immediate tension between the existence of hidden variables and quantum mechanics because of its complementary (noncommuting) variables.

Second, to be explicit, Bell and Mermin assume that the hidden variables do not require information exchange between spacelike separated events to do their work. In other words, since Alice and Bob can make their measurements simultaneously, we do not want Bob's measurement setting and/or outcome to influence Alice's measurement outcome (or vice versa). That would suggest an instantaneous (nonlocal/spooky) causal influence. Again, the hidden variables are being invoked precisely "to restore to the theory causality and locality" [10].

Finally, Bell and Mermin assume that the hidden variables are not correlated with the measurement settings (statistical independence). That simply means that the overall distribution of the eight instruction sets (RRR, RRG, RGR, GRR, GGG, GGR, GRG, RGG) produced by the source for all trials of the experiment must be the same as the distribution of the eight instructions for each of the nine different measurement setting choices (11, 22, 33, 12, 13, 21, 23, 31, 32). If all of these assumptions hold true, then we have the normal causation of Figure 3.3 and Bell's inequality (agreement in at least $\frac{1}{3}$ of all trials) will hold true.

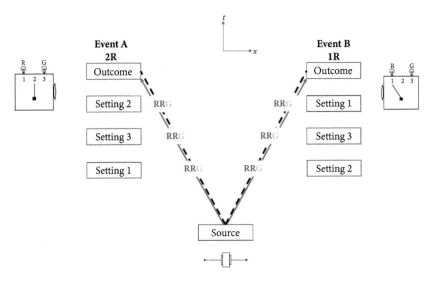

Fig. 3.3 Normal causation. Adding a vertical axis representing time t to Figure 2.1 with a spatial distribution in x gives this spacetime depiction of a trial for Mermin's device. Alice and Bob can change their detector settings right to the point of detection. The hidden variables (Mermin's instruction sets) represent some yet-to-be-determined dynamical law and/or mechanistic causal process between the particles and measurement equipment that produces the measurement outcomes. The particles are sent from a source, so the instruction sets are shown along the particle paths in spacetime from the source emission event to their respective measurement events (dashed lines). A causal account of the two measurement outcomes proceeds along the blue arrows.

These assumptions are all based on a constructive explanation of entanglement, so before we look closely at the consequences of violating any of these assumptions, let's explore the origin of this bias.

3.4 Conservation Principles, Dynamical Laws, and Entanglement

Since the mystery of entanglement as we have articulated it is 'average-only' conservation and Bell's theorem tells us that a constructive account of entanglement will entail the violation of locality, statistical independence, intersubjective agreement, and/or unique experimental outcomes, let's recall how conservation principles and dynamical laws are related. This will clarify exactly how the EPR and EPR–Bell paradoxes are based on a dynamical bias. To do that, let's go back to basic physics and look carefully at how the conservation principle for momentum is related to Newton's dynamical laws of motion. Of course, momentum is part of Newtonian dynamics (kinematics + mass → momentum, force, energy, etc.) in contrast to its kinematics (displacement, velocity, acceleration), so by "dynamical laws" here we mean in contrast to principle constraints, such as conservation principles (Table 3.1).

Typically, conservation of momentum is derived from Newton's laws, so that the dynamical law is viewed as more fundamental than the conservation principle. That is, we start with Newton's second law,

$$\sum_{i=1}^{N} \vec{F}_i = \frac{d\vec{p}}{dt}, \tag{3.1}$$

which simply says that the N forces acting on a bodily object give rise to its change of momentum, $\vec{p} = m\vec{v}$, where m is the mass of the bodily object and $\vec{v}$ is its velocity. If the forces on the bodily object sum to zero, then $d\vec{p}/dt = 0$, so $\vec{p}$ is constant, i.e., it is conserved. If this is the case, Noether's theorem tells us the dynamics for this bodily object in this particular situation (its action) is invariant with respect to spatial translations and we say the dynamics has a spatial translational symmetry (recall, momentum generates spatial translations). While it is not common to do so, we can flip the explanation so the conservation principle gives rise to the dynamical laws with their symmetries for specific cases [25].

Table 3.1 Comparison of Newtonian and quantum-mechanical constraints and dynamical laws. Conservation of momentum, angular momentum, and energy each have a corresponding dynamical law in Newtonian mechanics. [We have included conservation of energy in this table, where W is work and U is potential energy, for the curious reader.] 'Average-only' conservation for the Bell states does not have a corresponding dynamical law in the formalism of quantum mechanics, since the conservation is contained in the Bell states themselves (as we will explain in Chapter 7), while the dynamics of quantum mechanics, i.e., Schrödinger's equation, governs the time evolution of the state vector among those states. If any such dynamical counterpart to conservation per the Bell states exists, it must be 'hidden'.

Constraint	Dynamical law(s)
Conservation of momentum	$\vec{F} = \dfrac{d\vec{p}}{dt}$
Conservation of angular momentum	$\vec{\tau} = \dfrac{d\vec{L}}{dt}$
Conservation of energy	$dW = \vec{F} \cdot d\vec{x}$ and $\vec{F} = -\vec{\nabla}U$
'Average-only' conservation	None

That is, we start by assuming the (global) conservation of momentum for all the bodily objects in the universe,

$$\sum_{i=1}^{n} \vec{p}_i = \text{constant},\qquad(3.2)$$

where the number of bodily objects in the universe n would be a very large number, of course. In practice, we only deal with small subsets under the assumption that these subsets are isolated from (do not exchange momentum with) all the other bodily objects in the universe to within experimental limits. That is typically how one uses Eq. (3.1), for example. Equation (3.2) means

$$\sum_{i=1}^{n} \frac{d\vec{p}_i}{dt} = 0.\qquad(3.3)$$

Let us focus on the nth bodily object, rewriting Eq. (3.3) as

$$\frac{d\vec{p}_n}{dt} = -\sum_{i=1}^{n-1} \frac{d\vec{p}_i}{dt} \equiv \sum_{i=1}^{N} \vec{F}_i.\qquad(3.4)$$

Equation (3.4) is Newton's second law [Eq. (3.1)], as well as Newton's third law (for every action there is an equal and opposite reaction), while Eq. (3.2) is Newton's first law (an object in motion tends to remain in motion and an object at rest tends to remain at rest). The two approaches are mathematically equivalent, but most people consider the dynamical approach (with Noether's theorem applied to isolated systems) to be more fundamental than the global conservation of momentum because it is more 'causal'. This is the dynamical/constructive bias in terms of Newtonian mechanics, so let's look closely at where that originates.

When the momentum of bodily object X changes, we want to know why. We do not typically explain its change in momentum by relating it to the change in momentum for other bodily objects; instead, we say the other bodily objects exerted a force on bodily object X. That force is the causal mechanism that acted locally on bodily object X to change its momentum. For example, to explain why the window is broken, we would say Timmy exerted a force on the rock (by throwing it) and the rock then exerted a force on the glass (by colliding with it). We would not say the rock and Timmy exchanged momentum then the glass and rock exchanged momentum.

We want to explain events in terms of cause and effect whenever possible and to say Timmy and the rock exchanged momentum loses a key element in the causal explanation, i.e., Timmy threw the rock. If Timmy had been hit by the rock, it would also be the case that Timmy and the rock exchanged momentum, but that situation is very different from Timmy throwing the rock. For explanatory purposes, it matters causally whether or not Timmy is exerting a force on the rock or the rock is exerting a force on Timmy.

Of course, Newton's third law says the force exerted by Timmy on the rock is equal and opposite to the force exerted by the rock on Timmy, regardless of the context. But, our decision about which bodily object is having its momentum changed by a force and which is the origin of that force depends on the causal context.

Obviously, there are situations when using a conservation principle rather than a dynamical law makes sense. For example, why does an ice skater spin faster after pulling her arms tightly into her body? Conservation of angular momentum. We do not say that the various subsets of her body exerted equal and opposite torques on each other such that, for example, her shoulders caused her arms to speed up and vice versa. But, we could explain the phenomenon using dynamical laws and we tend to believe the origin of that conservation principle is dynamical, so the conservation principle is not fundamental.

That is, just like conservation of momentum is derived from Eq. (3.1), one typically starts with

$$\sum_{i=1}^{N} \vec{\tau}_i = \frac{d\vec{L}}{dt} \tag{3.5}$$

to derive conservation of angular momentum. This equation is simply saying that the N torques acting on a bodily object give rise to its change of angular momentum, $\vec{L} = I\vec{\omega}$, where I is the moment of inertia for the bodily object and $\vec{\omega}$ is its angular velocity. As with momentum, if the torques sum to zero, then $d\vec{L}/dt = 0$, so $\vec{L}$ is constant, i.e., it is conserved. Noether's theorem then tells us the dynamics for this bodily object in this particular situation (its action) has a spatial rotational symmetry (recall, angular momentum generates spatial rotations). Naturally, we could also replicate Eqs. (3.2)–(3.4) for angular momentum, leaving us with the same question: which approach is more fundamental, the dynamical law or the conservation principle?

Again, the answer would be that most people consider Eq. (3.5) (with Noether's theorem applied to isolated systems) to be more fundamental than the conservation of angular momentum globally because torque is locally causal,[2] exactly like force. Even if one is using the conservation of angular momentum to explain an observation, e.g., the skater's changing angular velocity, one could always explain the phenomenon at a more fundamental level using equal and opposite torques. That would be the 'real' cause of the skater's increased rate of rotation, after all she can *feel herself* exerting that torque between her shoulders and arms. There is another key difference between these two types of explanation that we can see using our objective spacetime model of reality.

That is, a conservation principle is used to relate the values of the conserved quantity for the bodily object at two distinct Times (Initial Time and Final Time), i.e., on two NOW slices of Space separated by finite (not infinitesimal) Time,[3] while a dynamical law is applied to the bodily object in just one NOW slice of Space.[4] Recall the subjective spacetime model of reality for you at home getting a text message from your friend at the beach (Figure 1.5). That

---

[2] Do not confuse this notion of local with locality. The terms are related because when we talk about locality we need a pair of spacelike-related events, and those events exist locally in spacetime [3]. So, when we talk about locality we are talking about causes that apply locally in spacetime without spacelike influence.

[3] Think of how you used conservation principles to solve problems in basic physics.

[4] Remember free-body diagrams in basic physics?

subjective spacetime model of reality would contribute to the objective space-time model of reality in Newtonian physics because $c = \infty$ (as we will see in Chapter 4).

In the Newtonian objective spacetime model of reality, all subjective space-time models of reality share the same NOW slices. In other words, what you say exists NOW in the conversation with your friend at the beach agrees with what they say exists NOW. The events and bodily objects in the NOW slices of their subjective spacetime model of reality would be the same as yours. And, the values of Time for each NOW slice could be made the same for both subjective spacetime models of reality in accord with Newton's absolute space and absolute time. So, all observers would agree as to what events came into existence together and what events stopped existing together, instant by instant.

In the Newtonian view, everyone tells the same story about events being produced by dynamical laws acting on bodily objects located in Space at some particular Time. Applying this to the ice skater, we see that she sped up between the Initial Time and the Final Time because of the equal and opposite torques acting within her body instant by instant between those two Times. So, while both the conservation principle and dynamical law explain what happened to the ice skater, the dynamical law is the more basic of the two explanations because it is a causal mechanism applied instant by instant in continuous fashion through Time and locally in Space. That is, forces and torques are causal mechanisms that explain why certain events come into existence on a NOW slice, i.e., forces and torques produce new events, so forces and torques with Noether's theorem explain conservation principles, not the converse.

This causal understanding of reality (captured in Newton's objective space-time model of reality) has two immediate benefits. First, by identifying dynamical laws we can manipulate and control reality to create new technology to improve the quality of our lives, e.g., internal combustion engine vehicles. Second, by identifying dynamical laws we can explain the past in terms of counterfactual actions in order to avoid undesirable outcomes in the future.

For example, suppose you rounded a bend in your internal combustion engine vehicle (car) and slid off the road, causing damage to your car. The causal explanation of that car wreck allows you to identify the following counterfactual: If you had reduced your speed on that curve in the road, the static frictional force between your tires and the road would not have been exceeded. In that counterfactual case, you would not have slid off the road, so you now have ways to avoid that particular unpleasant consequence in the future, i.e., slow down when driving around curves and/or increase the coefficient of static friction between your tires and the road.

To summarize the constructive perspective above, forces and torques cause changes in the momenta and angular momenta of bodily objects, respectively. Forces and torques are causal mechanisms that produce new events on each NOW slice of Space per the Newtonian objective spacetime model of reality. This causation is unambiguous because all observers agree on which events are coming into existence and which events are going out of existence instant by instant, i.e., their subjective spacetime models of reality have the same NOW slices of Space as the (intersubjectively agreed upon) Newtonian objective spacetime model of reality. Since this maps beautifully to each person's subjective dynamical experience per our everyday interactions with each other, constructive explanation is the preferred means of explanation. As Valia Allori wrote [6]:

> Philosophers of science have proposed objective accounts of explanation, but they all recognize there's a strong sense in which explanation is "explanation for us," and any account should capture our intuition that explanation is fundamentally dynamical. This is connected with causation: intuitively, we explain an event because we find its causes; causes happen before their effects and "bring them about."

When forces and torques are nonexistent or sum to zero for some system, that isolated system's momentum and angular momentum do not change, i.e., momentum and angular momentum are conserved for the isolated system. Accordingly, Noether's theorem says the dynamics for that isolated system (its action) has a spatial translational and rotational symmetry. If two (or more) parts of an isolated system are spatially distinct, one can still use conservation principles to account for the correlated behavior of its spatially separated parts, e.g., Bell state conservation explaining Facts 1 and 2 for cases (a) and (b) for the Mermin device. However, given what we know about dynamical laws and conservation principles above, it would seem that a more basic constructive explanation should be sought to explain Bell state conservation, e.g., hidden variables acting locally in space via dynamical laws in accord with locality and statistical independence.

As EPR pointed out, quantum mechanics fails this expectation concerning Fact 1 for case (a) because its formalism says nothing about the hidden variables, i.e., it says nothing about the dynamical law(s) responsible for Bell state conservation (Table 3.1). Without the hidden variables, one is left with a violation of locality, statistical independence, intersubjective agreement, and/or unique experimental outcomes to account dynamically for the exact conservation of spin angular momentum associated with Fact 1 for case (a). That is the EPR paradox.

Bell then pointed out that hidden variables acting in accord with locality and statistical independence cannot account for the 'average-only' conservation of spin angular momentum associated with Fact 2 for case (b). Thus, we do not have an acceptable dynamical law for the 'average-only' Bell state conservation of spin angular momentum for the Mermin device. That is the EPR–Bell paradox (Table 3.1).

This (heuristic) overview shows how these paradoxes are rooted in a particular dynamical bias as represented by Newtonian mechanics. As you can see, it is not an unreasonable bias given the success of Newtonian mechanics and the fact that quantum mechanics gives us Newtonian mechanics on average. Now let's look at the (selectively abbreviated) history of constructive models that violate Bell's inequality in accord with the quantum-mechanical prediction to appreciate what the violations of locality, statistical independence, intersubjective agreement, or unique experimental outcomes entail.

3.5 Superluminal Causation

One of the earliest interpretations of quantum mechanics called the de Broglie–Bohm interpretation assumes the existence of an instantaneous causal influence called the *pilot wave*. It was developed by Louis de Broglie in the 1920s and revived by David Bohm in the 1950s. Accordingly, the pilot wave instantly updates the universe with information concerning Bob's (Alice's) measurement on his (her) particle and its outcome. Such a causal influence could indeed account for Facts 1 and 2 as depicted in Figure 3.4.

There are generally two reasons instantaneous causation is not viewed favorably by physicists. First, there is a desire that causes precede their effects and instantaneous causation has no such temporal order. Of course, we can avoid that complaint by simply supposing the measurements are not *exactly* instantaneous, but Bob's (Alice's) measurement actually precedes Alice's (Bob's) measurement by some small time, so as to create a temporal order. In that case, the speed of the causal influence from Bob's (Alice's) detector to Alice's (Bob's) detector is very fast, but not infinite.

The problem with that explanation is that Einstein's theory of special relativity (scc Chapter 4) says the temporal order of two events related by such

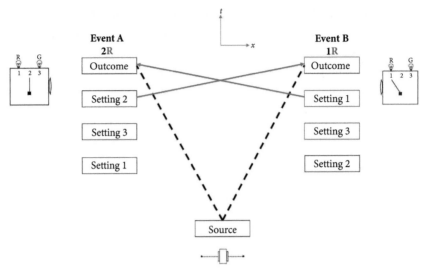

Fig. 3.4 Superluminal causation. The hidden variables require a spacelike exchange of information between measurement settings and detectors. In other words, the correlation between events A and B is explained by A causing B or B causing A (blue arrows) per Reichenbach's Principle. Since events A and B are correlated by faster-than-light causation, this is what Einstein derided as "spooky actions at a distance."

faster-than-light influences is ambiguous. That is, if Bob and Alice's measurement events are related in spacelike fashion, then Bob's measurement event would occur before Alice's in some reference frames while Alice's measurement event would occur before Bob's in other reference frames. Indeed, there would even be a reference frame in which Bob and Alice's measurement events are simultaneous, again wiping out any temporal order to cause and effect. So as to have all reference frames agree on the temporal order of two events, the two events must be related by slower-than-light influences, i.e., in timelike fashion (Figure 1.2). This invariant causal structure of timelike related events in special relativity's spacetime is highly regarded in physics, and spacelike causation is precisely what Einstein deemed "spooky." In a letter to Paul Epstein in 1945, Einstein says such spacelike causation [38, p. 144]

is of course logically possible, but it is so very repugnant to my physical instinct that I am not in a position to take it seriously—entirely apart from the fact that we cannot form any clear idea of the structure of such a process.

Given Einstein's pejorative characterization of spacelike causation, we might expect him not to advocate the de Broglie–Bohm interpretation. Indeed, in 1952 he wrote to Born [13]:

Have you noticed that Bohm believes (as de Broglie did, by the way, 25 years ago) that he is able to interpret the quantum theory in deterministic terms? That way seems too cheap to me. But you, of course, can judge this better than I.

Even de Broglie–Bohm advocate Aephraim Steinberg conceded [21]:

The universe seems to like talking to itself faster than the speed of light. I could understand a universe where nothing can go faster than light, but a universe where the internal workings operate faster than light, and yet we're forbidden from ever making use of that at the macroscopic level— it's very hard to understand.

What he means by "making use of that at the macroscopic level" is that Alice and Bob cannot send information to each other faster than light using measurements on entangled particles. That is called *no signaling* and we will say a bit more on that in Chapter 8. Steinberg's point is that whatever superluminal causal influence the universe is using to violate Bell's inequality with microscopic entangled particles is not available for us to exploit at the macroscopic level and Steinberg finds that "hard to understand."

Some advocates of the de Broglie–Bohm interpretation argue that there is actually a 'true' reference frame in which one measurement did influence the measurement outcome of its entangled partner in spacelike fashion. Accordingly, the observers in all the other reference frames are being fooled by the temporal order of their observations. As we explained in Chapter 1, the idea that there is a preferred reference frame was also invoked in the late 1800s, i.e., the rest frame of the luminiferous aether. Meter sticks moving through the aether would shrink (length contraction) in just the right way so that observers in all reference frames in motion through the aether would measure the same value for the speed of light c, regardless of their motion relative to the source (light postulate, an empirically discovered fact).

As we will see in Chapter 4, the preferred reference frame of the aether was gradually abandoned after Einstein introduced the theory of special relativity in 1905. Therein, Einstein rejected the preferred reference frame of the aether as a constructive explanation for length contraction, which people were using to explain the light postulate. Instead, he flipped the entire explanatory sequence by invoking the relativity principle, i.e., all inertial reference frames are equally valid.

Given what happened with special relativity, physicists are loath to invoke a preferred reference frame to solve the mystery of entanglement. For example, Bell stated [18, pp. 48–49]:

> The cheapest resolution [of the puzzle of nonlocality] is something like going back to relativity as it was before Einstein, when people like Lorentz and Poincaré thought that there was an aether.

Nonetheless, Bell was an advocate of Bohmian mechanics and said that it should be taught in courses on quantum mechanics. He felt that we just needed to find a Lorentz-invariant version so as to avoid a conflict with special relativity [12]:

> For me this is the real problem with quantum theory: the apparently essential conflict between any sharp formulation and fundamental relativity.

Others have also complained that entanglement seems to put quantum mechanics at odds with special relativity. Sandu Popescu and Daniel Rohrlich write [36]:

> Quantum mechanics, which does not allow us to transmit signals faster than light, preserves relativistic causality. But quantum mechanics does not always allow us to consider distant systems as separate, as Einstein assumed. The failure of Einstein separability violates, not the letter, but the spirit of special relativity, and left many physicists (including Bell) deeply unsettled.

Indeed, as we stated previously, any constructive explanation of the quantum-mechanical violations of Bell's inequality will involve spacelike causation (nonlocality) in violation of the invariant causal structure of special relativity (assuming statistical independence, intersubjective agreement, and unique experimental outcomes). Since most physicists are only considering constructive explanations of entanglement and do not consider a violation of statistical independence, intersubjective agreement, or unique experimental outcomes to be viable options, their belief that entanglement entails a conflict between quantum mechanics and special relativity is reasonable.

But, quantum mechanics is just a low-energy approximation to quantum field theory and quantum field theory *is* Lorentz invariant. So if the particle energies in your experiment are low enough that the predictions of quantum mechanics hold, then your experiment is in accord with Lorentz-invariant quantum field theory. And, entanglement is one of the many predictions of quantum mechanics to be experimentally verified and technologically exploited. Therefore, entanglement is totally compatible with Lorentz invariance and cannot be at odds with special relativity in that respect.

Of course, that hasn't stopped people from working on other Lorentz-invariant accounts of the violation of Bell's inequality and Maudlin has studied these concluding [31, p. 202]:

> the common thread that runs through all of these proposals is that no results are to be had at a low price. Indeed, the cost exacted by those theories which retain Lorentz invariance is so high that one might rationally prefer to reject Relativity as the ultimate account of space-time structure.

So, there is absolutely no reason to believe we need a Lorentz-invariant account of the violation of Bell's inequality to replace the one we already have (quantum mechanics from quantum field theory), and attempts to construct such an alternative have produced cures worse than the perceived disease.

However, given relativistic quantum field theory, while it is the case that quantum mechanics (and therefore the violation of Bell's inequality) is quite compatible with special relativity, you can produce an artificial incompatibility. For example, if you invoke superluminal causality as in Bohmian mechanics to explain the violation of Bell's inequality, then you need a preferred reference frame, so you have violated the invariant causal structure of special relativity's spacetime, at least in spirit. Again, this invariant structure is one of the most highly regarded consequences of special relativity, so physicists on the whole have not subscribed to constructive accounts like Bohmian mechanics.

The bottom line is that both special relativity and quantum mechanics map beautifully to experiments where the theories are expected to be applicable, so there is no experimental evidence to suggest that special relativity and quantum mechanics are somehow at odds, and that includes entanglement. Indeed, given that experiments verifying quantum mechanics are also verifying the low-energy version of Lorentz-invariant quantum field theory, we would expect to discover how special relativity and quantum mechanics are *compatible* in some deep sense. We will do just that in Chapters 6 and 7 where we show that both theories are based on the invariant measurement of a fundamental constant of Nature in accord with the relativity principle (the speed of light c for special relativity and Planck's constant h for quantum mechanics).

By invoking superluminal causation, Bohmian mechanics abandons the locality assumption of Bell's theorem. Is there some way to keep the instruction sets (realism) and locality and still reproduce Facts 1 and 2 of the Mermin device? Yes, of course, we can abandon statistical independence.

3.6 Retrocausality

One constructive way to account for Facts 1 and 2 without violating locality was proposed in 1947 (many years before Mermin's paper) by Olivier Costa de Beauregard. Like Costa de Beauregard, some argue that all we need for our Bell-inequality-violating causal influence not to violate locality is that it act in timelike fashion, and we can do that by allowing the causal influence to travel *backward* in time (Figure 3.5). This is retrocausality and it solves the mystery of entanglement by abandoning Mermin and Bell's assumption that the particles do not 'know' what measurement settings they will encounter when they leave the source.

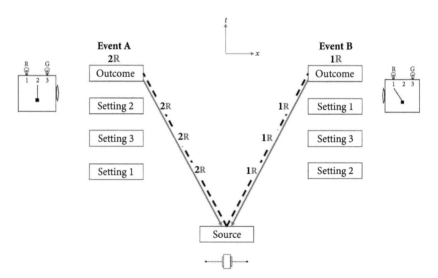

Fig. 3.5 Retrocausality. The correlation between events A and B is explained by a causal influence from events A and B into their past to the source emission event (blue arrows). Accordingly, a timelike causal solution to the mystery of entanglement is possible.

We should point out that retrocausality does not allow one to send messages (signals) into the past, so Ken Wharton and Nathan Argaman [42] use the term "future-input dependence" rather than "retrocausality." Argaman writes [8]: "Signaling into the past must thus indeed be strictly impossible, as it would allow construction of causal loops—the well-known inconsistency arguments of the grandfather paradox (a.k.a. the bilking argument)."

This is a violation of statistical independence because the hidden variables are certainly correlated with the measurement settings since they are actually determined by the measurement settings. Again thinking in terms of superluminal causality, Costa de Beauregard wrote [22]:

> Einstein of course is right in seeing an incompatibility between his special relativity theory and the distant quantal correlations, *but only under the assumption that advanced actions are excluded.*

The term "advanced actions" in his interpretation was inspired by the work of Wheeler and Feynman.

In 1945, Wheeler and Feynman pointed out that Maxwell's equations for light (electromagnetic) waves allow solutions that move forward in time from the source to the absorber (called retarded waves), as well as solutions that move backward in time from the absorber to the source (called advanced waves). Physicists generally ignore the advanced wave solutions, believing them to be a mere mathematical anomaly, but Costa de Beauregard suggested taking them seriously for quantum mechanics because an advanced wave solution is actually used in quantum mechanics.

That is, in quantum mechanics one solves Schrödinger's equation for the wavefunction[5] $\psi(x, t)$, which is used to compute the probability of finding the particle near position x at some time t. One obtains the probability (per unit length) from the wavefunction by multiplying $\psi(x, t)$ by its complex conjugate $\psi^*(x, t)$. As it turns out, $\psi^*(x, t)$ can be thought of as the time reversal of $\psi(x, t)$, i.e., an advanced wave of sorts, so one can understand the probability function as being composed of a part that moves forward in time from the source, $\psi(x, t)$, multiplied by a part that moves backward in time from the absorber, $\psi^*(x, t)$. Adlam calls causality that is "mediated by physical systems travelling backwards in time" *dynamical retrocausation* [1].

Specific instantiations of dynamical retrocausality include Yakir Aharonov, Peter Bergmann, and Joel Lebowitz's two-states vector formalism introduced

---

[5] This is the wavefunction in space and time, not the vector version in Hilbert space.

in 1964 [5] and John Cramer's transactional interpretation introduced in 1986 [16]. However, Cramer admits the backward-causal elements of his transactional interpretation are "only a pedagogical convention," and that in fact "the process is atemporal" [17]. That brings us to another way of thinking about retrocausality that Adlam calls *all-at-once retrocausality* [1]. Huw Price and Ken Wharton are staunch advocates for all-at-once retrocausality and argue that "the hypothesis that a quantum world of the kind Einstein hoped for would need to be retrocausal" [37]. We will discuss that view in more detail below.

As stated earlier, the problem most physicists have with dynamical retrocausality is that causes do not precede their effects. Sabine Hossenfelder states[6] [28]:

> This behavior is sometimes referred to as "retrocausal" rather than superdeterministic, but I have refused and will continue to refuse using this term because the idea of a cause propagating back in time is meaningless.

That is, we generally think of a cause as bringing its effect into existence as described in Section 3.4, e.g., Timmy throws a rock at $t = 0$ breaking a window at $t = 1$ s. The act of throwing the rock at $t = 0$ brings the broken window at $t = 1$ s into existence. If Timmy had not thrown the rock at $t = 0$, the window would not be broken at $t = 1$ s. But, suppose causal influences propagate from future events backward in time to contribute causally to the production of present events, then intuitively that is difficult to square with the belief that future events have never existed. How can a causal influence that dynamically produces events in the current NOW slice of Space emanate from an event that has never existed? That is certainly not in accord with the Newtonian view of dynamical causation. Adlam writes [1]:

> Newton bequeathed to us a picture of physics in which the fundamental role of laws is to give rise to time evolution: the Newtonian universe can be regarded as something like a computer which takes in an initial state and evolves it forward in time.

Consequently, opponents of dynamical retrocausality complain that it is not really causal at all. For example, Ruth Kastner writes [29]:

---

[6] She says "'retrocausal' rather than superdeterministic" because she defines superdeterminism more broadly than we do here. For her, any violation of statistical independence is superdeterminism.

Time-symmetric interpretations of quantum theory are often presented as featuring "retrocausal" effects in addition to the usual forward notion of causation. This paper examines the ontological implications of certain time-symmetric theories, and finds that no dynamical notion of causation applies to them, either forward or backward. It is concluded that such theories actually describe a static picture, in which the notion of causation is relegated to a descriptor of static relationships among events.

All-at-once retrocausality is what Kastner describes as "a descriptor of static relationships among events." Wharton and Raylor Liu write [43]:

> Such models generally result from taking seriously the block universe view of general relativity, such that any given parameter at any given location in spacetime can only have one value. By allowing both initial and final inputs into the calculation, but only permitting one meaningful result at every intermediate spacetime location, the entire problem must be solved 'all at once', denying any objective dynamical or causal 'flow' from one time to another. Such models are still accurately termed 'retrocausal', because the future inputs are constraining past parameters ...

Adlam writes [1]:

> For example, the solution to the Einstein equations of General Relativity is not a state at a time but an entire history of a universe ... so it doesn't seem to require any concept of time evolution. A time-evolution formulation of the Einstein equations does exist ... but the original global formulation remains central to research in the field and there seems no obvious reason to think that the time evolution formulation must be more fundamental.

This *all-at-once* or *adynamical/acausal global constraint (AGC)* view is "governed by the global topology of the connections through space and time"[7] [43] and is consistent with the objective spacetime model of reality in special relativity, which holds in small (flat) regions of the globally curved spacetime for general relativity just as one can assume small regions on the curved surface of Earth are flat (more on that in Chapter 9). That is, all observers agree on what events coexist (what events constitute a particular NOW slice of Space) in the Newtonian objective spacetime model of reality, but that is not the case in the objective spacetime model of reality for special relativity.

---

[7] Topology is the study of shapes without regard to their size.

Events that coexist are said to come into existence simultaneously in accord with our intuition (Section 3.4). In the Newtonian objective spacetime model of reality, if Events 1 and 2 coexist (happen simultaneously) for Alice, then they coexist (happen simultaneously) for Bob, regardless of how Alice and Bob's inertial reference frames are related. In special relativity's objective spacetime model of reality, it is possible for Events 1 and 2 to coexist for Alice while Bob says Event 1 happened before Event 2 and Charlie says Event 1 happened after Event 2 when Alice, Bob, and Charlie reside in three different inertial reference frames. This ambiguity in the temporal ordering of spacelike related events can be problematic for dynamical retrocausality, but it is quite compatible with an all-at-once view.

For example, there is an experimental configuration called the *triangle-network experiment*[8] for which [43]:

> The role of the future settings is conspicuously absent; no adjustable settings are even required to prove the no-go theorem of the triangle network. Instead, the non-classical behavior seems to be governed by the global topology of the connections through space and time; it is this triangle pattern [in spacetime] which must somehow create the unusual correlations.

That means there are no future boundary conditions to "retro-time evolve" for dynamical retrocausality in the triangle-network experiment. So, of the two retrocausal approaches the all-at-once version seems preferable to the dynamical version. Adlam writes [1]:

> Developments in modern physics give us good reason to take seriously the possibility of laws which are non-local, global, atemporal, retrocausal, and/or not easily put in the standard kinematical-dynamical form [of Newtonian mechanics].

The bottom line is, for those who want a constructively causal solution to the mystery of entanglement that does not violate locality, dynamical retrocausality just does not cut it. However, if one is willing to abandon dynamical causality, all-at-once retrocausality without hidden variables is viable and we will discuss that further in Chapter 5. So, is there some way to keep locality, realism, Facts 1 and 2 of the Mermin device, *and* a sense of causal becoming? Again, yes, but it might be the most 'desperate' move yet.

---

[8] The details of this experiment are not necessary for making our point, but you can read [43] if you would like to see the details.

3.7 Superdeterminism and Supermeasured Theory

In a 1985 BBC radio interview with Paul Davies, Bell himself said [18]:

> There is a way to escape the inference of superluminal speeds and spooky action at a distance. But it involves absolute determinism in the universe, the complete absence of free will. Suppose the world is super-deterministic, with not just inanimate nature running on behind-the-scenes clockwork, but with our behavior, including our belief that we are free to choose to do one experiment rather than another, absolutely predetermined, including the 'decision' by the experimenter to carry out one set of measurements rather than another, the difficulty disappears.

Again, the assumption we are giving up with superdeterminism is statistical independence, i.e., that Alice and Bob's detector settings are not correlated with the hidden variables. Mermin was careful to make this assumption explicit by saying each instruction set is measured with equal frequency in all nine detector setting pairs. So, over the course of the numerous trials of the experiment, each instruction set is measured in detector setting 11 as often as 12 as often as 13 as often as 21, etc. If we relax this assumption, it is possible to reproduce Facts 1 and 2 for the Mermin device without Alice and Bob being any the wiser. Let's see how that works.

In row 2, column 2 of Table 3.2, you can see that Alice and Bob select (by whatever means) setting pairs 23 and 32 with twice the frequency of 21, 12, 31, and 13 in those case (b) trials where the source emits particles with the instruction set RRG or GGR (produced with equal frequency). Column 4 then shows that this hidden 'conspiracy' would indeed satisfy Fact 2. However, the detector setting pairs would not occur with equal frequency overall in the experiment and this would certainly raise red flags for Alice and Bob. There-fore, we introduce a similar disparity in the frequency of the detector setting

Table 3.2 Superdeterminism and the Mermin device.

Instruction Set	Doubled Measurements in Blue	Case (a) Results	Case (b) Results
RRG GGR	11 12 13 21 22 23 31 32 33	RR RR RR GG GG GG GG RR	RR RG RR RG RG GR GR GR GG GR GG GR GR RG RG RG
RGR GRG	11 12 13 21 22 23 31 32 33	RR GG GG RR GG RR RR GG	RG RG RR GR GR GR RR RG GR GR GG RG RG RG GG GR
RGG GRR	11 12 13 21 22 23 31 32 33	RR GG GG GG GG RR RR RR	RG RG RG GR GG GR GR GG GR GR GR RG RR RG RG RR
RRR GGG	Not Created/Measured	—	—

pair measurements for RGR/GRG (12 and 21 frequencies doubled, row 3) and RGG/GRR (13 and 31 frequencies doubled, row 4), so that they also satisfy Fact 2 (column 4). Now, if these six instruction sets are produced with equal frequency, then the six case (b) detector setting pairs will occur with equal frequency overall.

In order to have an equal frequency of occurrence for all nine detector setting pairs, let detector setting pair 11 occur with twice the frequency of 22 and 33 for RRG/GGR (row 2), detector setting pair 22 occur with twice the frequency of 11 and 33 for RGR/GRG (row 3), and detector setting pair 33 occur with twice the frequency of 22 and 11 for RGG/GRR (row 4). Then, we will have accounted for Fact 1 (column 3) and Fact 2 (column 4) of the Mermin device with all nine detector setting pairs occurring with equal frequency overall.

Since the instruction set in each trial of the experiment cannot be known by Alice and Bob, they do not suspect any violation of statistical independence. That is, they faithfully reproduced the same quantum state in each trial of the experiment and made their individual measurements randomly and independently, so that measurement outcomes for each detector setting pair represent roughly $\frac{1}{9}$ of all the data. Indeed, Alice and Bob would say their experiment obeyed statistical independence, i.e., there is no (visible) correlation between what the source produced in each trial and how Alice and Bob chose to make their measurement in each trial.

We can imagine explaining this odd situation via some dynamical law or mechanistic causal process starting with some event in the past of the source emission event and the detector setting events (Figure 3.6). In other words, the hypothetical dynamical law or mechanistic causal process brings each emission event and the associated detector setting events into existence in just the right way to reproduce Table 3.2.

Advocates for superdeterminism include Nobel Laureate Gerard 't Hooft. What they like about superdeterminism is that you can keep so much of normal causal explanation. As Tim Andersen says, superdeterminism provides a "local, realist, reductionist interpretation with no extra dimensions, universes, or solipsism" [7], i.e., without violating locality, unique experimental outcomes, or intersubjective agreement. Consequently, they believe Einstein would definitely be an advocate of superdeterminism, especially given the constructive alternatives. But, of the constructive assumptions leading to Bell's theorem, none is more cherished by physicists than statistical independence. Why?

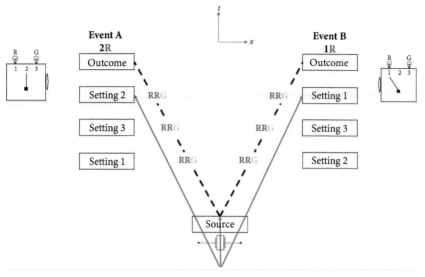

Fig. 3.6 Superdeterminism. The correlation between events A and B is explained by a causal influence from the past of event A, event B, and the source emission event (blue arrows). This causal influence does three things: it makes the source create the instruction set RRG, it makes Alice choose detector setting 2, and it makes Bob choose detector setting 1. According to superdeterminism, quantum mechanics is very incomplete and its completion would be highly nontrivial. For example, one would need an explanation of Alice and Bob's conscious decisions in order to fully understand entanglement.

The two main concerns about superdeterminism are the complexity of the unknown dynamical law or mechanistic causal process and the threat to the scientific method. The complexity problem is easy to appreciate because whatever the dynamical law or mechanistic causal process for this 'conspiracy' between detector settings and particle properties, it must govern any possible method for choosing detector settings. If Alice and Bob select detector settings using monkeys throwing darts, the Swiss lottery, or their own personal choices, no matter what the method is for choosing detector settings, this superdeterministic constructive explanation must account for it. Therefore, such a superdeterministic constructive explanation would be highly nontrivial. Bell writes [11]:

But this way of arranging quantum mechanical correlations would be even more mind-boggling than one in which causal chains go faster than light.

Apparently, separate parts of the world would be deeply and conspiratorially entangled, and our apparent free will would be entangled with them.

It is his last point that concerns most physicists about superdeterminism, i.e., its threat to the scientific method. As Zeilinger complained [44]:

When I do physics, I believe I can ask free questions of Nature to find out something new. If Nature is controlling the questions I ask, Nature could totally fool us. It would be the end of science.

Here, "free will" and "free questions" just represent the ability to sample Nature in truly random fashion, so we do not need to debate what constitutes free will.

Rather, to appreciate Zeilinger's complaint that "Nature could totally fool us," one has only to realize that Alice and Bob have no clue about what is happening behind the scenes, as shown in Table 3.2. As far as Nature allows them to know, Alice and Bob have made their measurements randomly and reproduced the source faithfully in all trials of the experiment. If Table 3.2 is true, then Nature is not only conspiratorial, but insidiously so.

To avoid the complaint that Nature is controlling the questions we can ask by controlling the measurement choices in accord with the hidden instruction set produced by the source, Jonte Hance, Sabine Hossenfelder, and Tim Palmer proposed supermeasured theory [26]. Here, "measured" refers to the term *measure* used in probability theory that dictates the distribution of states in the probability space, so "measured" does not refer to measurement. Let's look at the Mermin device to see what that means "without the gobbledygook" (to borrow Hossenfelder's phrase).

Table 3.3 Supermeasured theory and the Mermin device. The first row contains the nine possible settings and each column contains the instruction sets for each setting for each of the 72 trials of the experiment.

11	12	13	21	22	23	31	32	33
RR	RR	RR	RR	RR	RR	RR	RR	RR
RR	GG	GG	GG	RR	GG	GG	GG	RR
RR	RG	RG	RG	RR	RG	RG	RG	RR
RR	RG	RG	RG	RR	RG	RG	RG	RR
GG	RG	RG	RG	GG	RG	RG	RG	GG
GG	GR	GR	GR	GG	GR	GR	GR	GG
GG	GR	GR	GR	GG	GR	GR	GR	GG
GG	GR	GR	GR	GG	GR	GR	GR	GG

For the Mermin device, they would assign instruction sets to each setting pair for each trial of the experiment (Table 3.3). This means the distribution of hidden variables can vary from one setting pair to another. Thinking of it dynamically, it is the instruction sets that are controlled by measurement choices rather than measurement choices being controlled by instruction sets as in Figure 3.5. However, Hance et al. prefer to think of superdeterminism and supermeasured theories adynamically [26]:

In the supermeasured models that we consider, the distribution of hidden variables is correlated with the detector settings at the time of measurement. The settings do not cause the distribution. We prefer to use find [sic] Adlam's terms—that superdeterministic/supermeasured theories apply an 'atemporal' or 'all-at-once' constraint—more apt and more useful.

As we saw above, this is precisely the attitude of Wharton and Price for retrocausality. Indeed, supermeasured theory could be viewed as a possible approach to all-at-once retrocausality, although Hance et al. write, "we are not aware of an unambiguous definition of the term 'retrocausal' and therefore do not want to use it."

Regardless of the terminology, there seems to be an emerging consensus between these camps that the way to solve the mystery of entanglement specifically and to understand quantum mechanics more generally is via all-at-once or adynamical global constraints in spacetime. This is precisely where Adlam and Rovelli have taken relational quantum mechanics recently.

3.8 Relational Quantum Mechanics

Relational quantum mechanics "is an interpretation of quantum mechanics based on the idea that quantum states describe not an absolute property of a system but rather a relationship between systems" [4]. We should note that relational quantum mechanics was first 'cooked up' as a solution to the measurement problem, and that its account of entanglement has been a moving target over time. This will become clear as we move forward. Also note that we are not concerned with a detailed exposition of Rovelli's relational quantum mechanics here (we will say more in Chapter 6 about it as an information-theoretic reconstruction), we only need to know how it solves the mystery of entanglement. And, as we pointed out in Chapter 1, the original method of solving the mystery of entanglement in relational quantum mechanics was to omit Einstein's third step in the practice of physics.

Recall, in that third step observers compare their personal (subjective) data obtained in the different reference frames of their data collection devices and use it to build an objective model of reality (a model based on intersubjective agreement). As we stated in Chapter 1, if you abandon intersubjective agreement, you are abandoning the concern over spacelike causation because you are abandoning any meaning to a common spatial distribution of entangled events. Let's clarify this with the following example using the Mermin device.

The spacetime diagram for one trial with the Mermin device with three observers is shown in Figure 3.7. The source of the Bell state in this trial resides in the spacetime region Λ and all three observers form a relation with the source in Λ as indicated on their three worldlines O_1, O_2, O_3. Consider events along O_1, since the other two are similar.

The second event shown on worldline O_1 for observer 1 is [A,N]. At this event, information about Alice's measurement A (setting and outcome) and its causal past in spacetime region N form relations with observer 1. The third (final) event shown on worldline O_1 for observer 1 is [B,M]. At this event, information about Bob's measurement B (setting and outcome) and its causal past in spacetime region M form relations with observer 1. As the experiment is repeated, observer 1 will discover Facts 1 and 2 about cases (a) and (b), but these correlations will not be spacelike related, as they only pertain to timelike

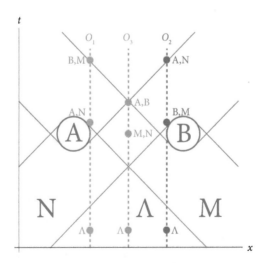

Fig. 3.7 Relational quantum mechanics and the Mermin device [35].

related events along O_1. Likewise for observers 2 and 3 along worldlines O_2 and O_3, respectively.

In order to create the EPR and EPR–Bell paradoxes, the observers would need to convene, consolidate their data into a single spacetime diagram (like that of Figure 3.7), and attempt to account for that data constructively. In Rovelli's original version of relational quantum mechanics, he avoids the EPR and EPR–Bell paradoxes by simply having each observer keep their correlated subjective relations to themselves.

Mermin adopts a similar stance [32]:

> My own science is at its root entirely about my own personal experience. The world that each of us infers from our own personal experience is often taken to be what science is about. Most of the time this is a hair-splitting distinction. But it is the failure to distinguish between my experience and the world that I infer from my experience that is the root of the confusion at the foundations of quantum mechanics. Whatever cannot ultimately be traced back to my personal experience cannot be a part of my science.

Mermin's view is called QBism [24], as developed by Chris Fuchs, Rüdiger Schack, and Carlton Caves. Accordingly, we each have our own personal science (physics here) that we use to establish expectations (via probability theory) about what we will experience when interacting with the world. "My physics" is built in part from the information I gather from my interactions with other physicists, and we do jointly construct spacetime diagrams housing our collective data like that of Figure 3.7. But, that is where the practice of physics stops for QBism, so there are no EPR and EPR–Bell paradoxes. Entanglement simply refers to the correlated events along each of our worldlines, and the paradoxes only arise when one tries to account for the collective data from all observers globally.

While the collectively produced Figure 3.7 does represent a weak form of intersubjective agreement, it lacks the level of coherence demanded by most physicists. In other words, the objective model of reality sought by most physicists would unify the disparate experiences shown along worldlines in Figure 3.7 by providing a common explanation for them, an explanation that applies to the collection as a whole.

Since this solution to the mystery of entanglement violates what most physicists consider germane to the practice of physics, it did not win consensus support in the foundations community. In response, Adlam and Rovelli modified relational quantum mechanics by adding Einstein's third step in the practice of physics, i.e., producing an objective spacetime model of reality. They write [4]:

> Thus in order for it be epistemically rational for us to believe RQM [relational quantum mechanics], it is necessary that there should be some mechanism for achieving intersubjective agreement between observers so we can have some idea of what the world is like for other observers. ... when an observer in RQM is involved in an interaction, the knowledge they obtain by looking at a measurement outcome must be recorded in their physical variables and must therefore be accessible to other observers. The accessibility of this knowledge will then guarantee that observers can align their perspectives by exchanging information, and therefore it will be possible to arrive at intersubjective agreement about the features of reality that we use to obtain empirical confirmation for scientific theories.

In doing so, they assumed an all-at-once or AGC ontology. As with Wharton, Price, Liu, Hossenfelder, Hance, and Palmer above, Adlam and Rovelli write [4]:

> However, the ontology we have suggested need not actually be understood in a time-asymmetric way. ... So there is no need to imagine that [sic] the set of quantum events being generated in some particular temporal order: they can simply be generated "all-at-once" in a time-symmetric and atemporal fashion.

We will have much more to say about this in Chapters 5 and 9, but we will point out here that this convergence towards an atemporal or all-at-once explanation of entanglement is in perfect accord with the constraint-based/structural explanation of Maltrana et al. [30], which is in stark contrast to *"causal mechanisms* that *explain mechanistically* the occurrence of a certain phenomenon" that we are calling constructive explanation. As pointed out by Adlam [2]:

> the existence of causal paradoxes is not a property of individual events but rather a holistic property of some collection of events, and therefore constraints forbidding them will not in general be expressible in local terms. That is to say, the constraints in question will probably have to be global "all-at-once" constraints, which as we have observed makes them a poor fit for a dynamic production picture.

Of course, the all-at-once constraint or AGC we are proposing is NPRF + h.

3.9 Many-Worlds

Many-Worlds is an old idea receiving renewed attention whereby all possible quantum outcomes are realized in every quantum interaction. Everything we said about relational quantum mechanics being first and foremost about the measurement problem also goes for Many-Worlds. While Many-Worlds is great fodder for science fiction, it does nothing to resolve the EPR–Bell paradox because entanglement is simply a fundamental property of reality in Many-Worlds (as we explained in Chapter 1). In fact, according to Many-Worlds it is the nonentangled classical reality of everyday experience that needs to be explained! This interpretation was created in response to the measurement problem, which arises for two reasons that Jeffrey Bub and Itamar Pitowski call "Two dogmas about quantum mechanics" [14]. These are important dogmas to address because they deal with constructive bias.

3.9.1 The Measurement Problem from the Constructive Bias

First "is John Bell's assertion that measurement should never be introduced as a primitive process in a fundamental mechanical theory like classical or quantum mechanics, but should always be open to a complete analysis, in principle, of how the individual outcomes come about dynamically" [14]. In classical mechanics, one solves the relevant differential or integral equation(s) to find, for example, position and velocity (which with mass gives momentum) for the bodily object(s) in question. Those properties as a function of time correspond directly to measurement outcomes for those variables at those times. That is, if the solution says the position of the bodily object at time t_1 is x_1, then a measurement of the position of the bodily object at time t_1 will produce an outcome of x_1. There is nothing additional corresponding to a measurement that needs to be added to the formalism of classical mechanics.

But in the differential equation formalism for quantum mechanics, one solves Schrödinger's (differential) equation and the solution gives the time evolution of the initial state (wavefunction) in the space of possible outcomes for some measurement (Hilbert space). The first dogma is violated because the wavefunction ($\psi(t)$) at any given time can (and usually does) represent more than one outcome for that particular measurement. In other words, if you do that particular measurement at time t_1, the outcome is not given by $\psi(t_1)$, unless $\psi(t_1)$ happens to be equal to only one of the possible measurement outcomes at that time (which is not usually the case). That means we need to add a "primitive process" to the differential equation formalism of quantum mechanics corresponding to a measurement to specify how one applies the formalism when a measurement is made.

Second "is the view that the quantum state has an ontological significance analogous to the significance of the classical state as the 'truthmaker' for propositions about the occurrence and non-occurrence of events, i.e., that the quantum state is a representation of physical reality" [14]. An interpretation of quantum mechanics in which the wavefunction has a counterpart in physical reality is called *psi-ontic*—psi for the wavefunction ψ and ontic from ontology for an element of reality.

From these two dogmas we have a measurement problem in this formalism because the method of time evolving the wavefunction in Hilbert space does not contain a mechanism to update the wavefunction after a measurement is made in violation of the first dogma. And, according to the second dogma, the wavefunction is an element of reality, so it should certainly be responsible for the dynamic production of (unique) quantum outcomes upon measurement, per our experience. You can see how these dogmas and the differential equation formalism of quantum mechanics map nicely onto our ignorance of the future and the idea that "*causal mechanisms ... explain mechanistically the occurrence of a certain phenomenon*" [30]. That is, the dogmas and this formalism are products of the constructive bias and create the measurement problem.

3.9.2 Solve or Avoid the Measurement Problem?

Of course, physicists are not confused about what constitutes a measurement in quantum or classical mechanics, they make measurements every day and apply the time-evolved quantum formalism to the outcomes. They simply add a postulate to the time-evolved formalism stating that the wavefunction changes ("collapses") into the state associated with whatever outcome is observed. We saw this in Section 3.2 when the quantum state being measured by the second measurement device was |R for setting 2⟩ and it became |G for setting 1⟩ after that second measurement. As Mermin points out [33], a collapse mechanism

> is also a feature of ordinary classical probability. When a statistician assigns probabilities to the answers to questions about a system, those probabilities vary in time by rules giving the smooth time evolution of the isolated

unquestioned system. But those probabilities also depend on any further information about the system from any other source. That updating of probabilities is the abrupt and discontinuous part of the classical process. Nobody has ever worried about a classical measurement problem.

But for those like Bell who subscribe to the two dogmas and the constructive bias, this collapse postulate is an ugly ad hoc feature of the time-evolved quantum formalism in need of a fix. Enter Hugh Everett III.

In his 1957 PhD thesis, Everett proposed a radical solution to the measurement problem, i.e., ignore the ad hoc measurement update altogether! Simply let the wavefunction continue to evolve according to Schrödinger's equation by combining the measurement device with the quantum system it measures. In that way, *all* of the possible outcomes of any experiment 'exist', each in its own new world in accord with the (physically real/ontic) wavefunction. In other words, if you do a particular measurement at time t_1, the outcome *is* given by $\psi(t_1)$, regardless of how many possibilities $\psi(t_1)$ represents. Here is how Philip Ball describes Many-Worlds [9]:

> It is the most extraordinary, alluring and thought-provoking of all the ways in which quantum mechanics has been interpreted. In its most familiar guise, the many-worlds interpretation suggests that we live in a near-infinity of universes, all superimposed in the same physical space but mutually isolated and evolving independently. In many of these universes there exist replicas of you and me, all but indistinguishable yet leading other lives.

As Ball points out, there are many problems created by Many-Worlds, so it was largely ignored until 1970 when Bryce DeWitt resurrected it in a *Physics Today* article [19]. While Many-Worlds solves the measurement problem, it is replaced by a new problem, i.e., the problem of finding 'you' in this process so as to obtain a frequency of outcomes [9]. This is what Lev Vaidman calls the "probability of self-location in a particular world" [41].

That is, return to our example in Chapter 1 with many trials of the quantum color measurement on a quantum glove and again suppose the quantum glove is prepared in a state of 25% White and 75% Black. In the collapse formalism, this means that the wavefunction will become White after measurement 25% of the time and it will become Black after measurement 75% of the time, which means that you will obtain a White outcome 25% of the time and a Black outcome 75% of the time. Now suppose there is no collapse, what do 'you' find to be true, given that both outcomes obtain after every measurement?

First, we put 'you' in scare quotes because there must be a version of you for each possible outcome. So, there would be one path through the repeated measurements in which 'you' obtained White every time and one path through the repeated measurements in which another version of 'you' obtained Black every time. How would those versions of 'you' know the wavefunction was 25% White and 75% Black? Indeed, those versions of 'you' would say the wavefunction was 100% White or 100% Black, and other versions of 'you' would obtain all the other possible distributions of White and Black. The measurement problem has been traded for what we might call the *probability problem*. Let's consider what this means for solving the mystery of entanglement per the Mermin device.

There, we must select combinations of Alice and Bob's worldlines through the extraordinarily complex maze of the Many Worlds such that along those worldlines, repeated experiments produce Facts 1 and 2 concerning cases (a) and (b), respectively. In order to solve this problem in constructive fashion, one needs to cook up a dynamical law or mechanistic causal process for the selection of these worldlines that can be applied in time-evolved fashion at each and every measurement event along each of Alice and Bob's worldlines. But, that is precisely the problem of solving the mystery of entanglement that we have using just one world.

Vaidman's answer to the probability problem is to simply postulate [41] "that the probability of self-location is proportional to the 'measure of existence' which is a counterpart of the Born rule of the collapse theories." That is, an admissible worldline for any 'you' through the maze of worlds is one that satisfies the quantum-mechanical probabilities for the quantum state in question. Many-Worlds simply defers to quantum mechanics as applied to one world using the collapse postulate to solve this problem. Essentially, Many-Worlds' answer to the probability problem relies (albeit indirectly) on the ad hoc collapse postulate it was designed to avoid in the first place. What is the implication of this for resolving the EPR and EPR–Bell paradoxes?

Suppose we have quantum gloves entangled in quantum color such that Alice obtains White 50% of the time and Black 50% of the time, and the same with Bob. Further, suppose that the entangled state says Bob and Alice must always obtain the same quantum color, so their paired outcomes in any given trial are never White–Black or Black–White. Why does that happen? EPR said we need nonlocality or hidden variables to explain that fact, that is the EPR paradox. Why does that happen according to Many-Worlds? Because the wavefunction says so.

That is, the world in which the experiment is being conducted splits into two worlds, one world in which both outcomes are White and one world in which both outcomes are Black, so there are no worlds in which one outcome is White and the other is Black. The nonlocality or need for hidden variables in the world of a given 'you' is understood to result from the nonlocality between different versions of 'you' in different worlds.[9] Vaidman says [41], "we should not disregard nonlocality of entanglement which requires the configuration space for its description." Paraphrasing Vaidman [39] for the Mermin device we would say:

> Specifying the Everett world of Alice fixes the world of Bob. This connection between local worlds of the observers is the nonlocality of [Many-Worlds].

That is, "spooky actions at a distance" act "between worlds" rather than within any given world according to the quantum state, but no further explanation is needed because quantum mechanics is fundamental. Now let's put all this together to see how Many-Worlds addresses the EPR–Bell paradox.

For the EPR–Bell situation, Alice and Bob make different measurements and can obtain mixed outcomes. Here, the mystery of entanglement is 'average-only' conservation (as shown in Chapter 0) in accord with the Bell-inequality-violating quantum-mechanical joint probabilities (shown in Chapter 2). So, to properly resolve the EPR–Bell paradox one needs to derive the quantum-mechanical joint probabilities without recourse to quantum mechanics. But, the probabilities in Many-Worlds are simply those of quantum mechanics obtained (essentially) by fiat, as explained above. Basically, according to Many-Worlds, the reason we see violations of Bell's inequality is because quantum mechanics predicts it and quantum mechanics is fundamental.

So, Many-Worlds solves the measurement problem, but creates the equally vexing probability problem and does nothing to solve the mystery of entanglement. Luckily, there is another formalism for quantum mechanics that can be paired with principle explanation to avoid the measurement problem altogether and solve the mystery of entanglement (as we show later). Let's look at this other formalism and see how it avoids the measurement problem.

The second formalism for quantum mechanics (called the path integral approach) simply computes the probability for each possibility directly, so there is no time-evolved wavefunction and therefore no ad hoc collapse postulate. Concerning this formalism, Wharton and Liu write [43], "the path integral

---

[9] The issue of nonlocality in Many-Worlds is complex and we cannot do it justice here. For further reading on this topic see [40, Section 7] and references therein.

assigns probabilities to entire patterns over space and time, not to instantaneous events." We will give some of the details in Chapter 9, but exactly how that is done does not concern us here; it suffices to know that the path integral formalism is most compatible with all-at-once, atemporal, or AGC explanation, which in turn is very compatible with principle explanation.

As we saw above and will see in Chapters 4 and 9, the objective spacetime model of reality according to special relativity and quantum mechanics makes much more sense via constraints in four-dimensional (4D) spacetime as opposed to dynamical laws and mechanistic causal processes for time-evolved states in 3D space (or Hilbert space). In the 4D view, one is asking about the distribution of entire spatiotemporal measurement configurations (to include specific outcomes) in 4D spacetime, while in the 3D time-evolved view one is asking about the distribution of outcomes in space and/or time independently.

The 4D view negates the first dogma because quantum events are not being brought into existence dynamically in the AGC explanation. And, it satisfies the first dogma in that the act of measurement is not being added as a "primitive process." Indeed, with the integral formalism and AGC explanation, measurement in quantum mechanics is as transparent as measurement in classical mechanics. That is, just as in the 3D time-evolved solution of the differential equation(s) of classical mechanics for (say) the position x and momentum p of bodily objects, one finds the values for $x(t_1)$ and $p(t_1)$ from the 4D solution of the integral equation(s) of classical mechanics in the 4D objective spacetime model of reality, and those values correspond directly to the corresponding measurement outcomes of position and momentum at time t_1. No measurement mystery here.

But, unlike the time-evolved solution $\psi(t)$ in Hilbert space from Schrödinger's equation, the path integral formalism of quantum mechanics *is* giving $\psi(t_1)$ for a specific measurement outcome in the 4D objective spacetime model of reality, i.e., no collapse needed. So, whatever you find for the distribution of quantum events for a particular quantum experiment—the relevant spatiotemporal context of source, mirrors, beam splitters, SG magnets, polarizers, etc. defining the quantum measurement—in the 4D objective spacetime model of reality corresponds directly to the quantum measurement outcome in question. Just like classical mechanics, no measurement mystery here. In short, the all-at-once perspective helps with the measurement problem just as it does with entanglement.

So, should we try to solve the measurement problem as with Many-Worlds? Or should we simply avoid the measurement problem by giving up the constructive bias that creates it? Given the problems for understanding measurement in quantum mechanics created by the constructive bias and the fact that Bell's theorem tells us any constructive solution to the mystery of entanglement must violate locality, statistical independence, intersubjective agreement, or unique experimental outcomes, we believe it is time to consider AGC principle explanation instead.

3.10 Summary

As we have seen in this chapter, physicists and philosophers have implemented increasingly 'desperate' and radical solutions to the mystery of entanglement to save constructive explanation in response to Bell's theorem. Since any constructive explanation of entanglement must violate locality, statistical independence, intersubjective agreement, and/or unique experimental outcomes according to Bell's theorem, it is not surprising that no constructive resolution of the EPR–Bell paradox has won consensus support in the foundations community. This explains why foundations is at an impasse.

A perfect example of this played out in a 2023 podcast conversation between Sean Carroll and Tim Maudlin. At one point, Maudlin stated [15]:

> What do I mean by explaining something? Well often I think it's a kind of causal explanation, I talk about cause and effect.

Later, when Carroll asked if Maudlin thought we would someday have a compelling answer to "Why the quantum?", Maudlin replied [15]:

> People like to think "Why the quantum?" ... you know ... you'll get this ... thing. I think oh come on just ... you know ... grow up ... no, I mean that there's gonna be foundations. You're going to hit a foundation, a plausible place to stop digging and all you can say is this is the way it is. Could it have been some other way? Yeah, it could've been some other way, but it isn't. If you're too pig-headed you're gonna get to the bottom and still be banging your head against it forever because there is nothing underneath it. I don't think you're going to be stopped by some master principle that says it could only have been this way. [Excerpted and condensed from his informal, conversational answer.]

Ironically, it is those who demand a constructive explanation of entanglement who have been 'banging their heads' against Bell's theorem and the violation of Bell's inequality. As we will see, there really is a "master principle that says it could only have been this way," i.e., NPRF + h.

We saw the same thing play out with the Ptolemaic system when epicycles and the equant were added to the deferent in order to save geocentricism and uniform circular motion in astronomy. We will see a similar situation in the next chapter concerning the luminiferous aether being used to save Galilean velocity transformation. As Chris Fuchs wrote [23]:

> Where present-day quantum-foundation studies have stagnated in the stream of history is not so unlike where the physics of length contraction and time dilation stood before Einstein's 1905 paper on special relativity.

In the next chapter we look at how Einstein dealt with that situation.

References

[1] E. Adlam, *Laws of Nature as Constraints*, 2021. Preprint. https://arxiv.org/abs/2109.13836.

[2] E. Adlam, *Two Roads to Retrocausality*, 2022. Preprint. https://arxiv.org/abs/2201.12934.

[3] E. Adlam, J. R. Hance, S. Hossenfelder, and T. N. Palmer, *Taxonomy for Physics Beyond Quantum Mechanics*, 2023. Preprint. https://arxiv.org/abs/2309.12293.

[4] E. Adlam and C. Rovelli, *Information is Physical: Cross-Perspective Links in Relational Quantum Mechanics*, 2022. Preprint. https://arxiv.org/abs/2203.13342.

[5] Y. Aharonov, P. Bergmann, and J. Lebowitz, *Time symmetry in the quantum process of measurement*, Physical Review, 134 (1964), pp. 1410–1416.

[6] V. Allori, *Book review of "Beyond the Dynamical Universe: Unifying Block Universe Physics and Time as Experienced," by Michael Silberstein, W. M. Stuckey, and Timothy McDevitt*, Metascience, 28 (2019), pp. 341–344.

[7] T. Andersen, *Superdeterminism may have solved the quantum measurement problem: A local, realist, reductionist interpretation with no extra dimensions, universes, or solipsism*, The Infinite Universe, 15 June (2021). https://medium.com/the-infinite-universe/superdeterminism-may-have-solved-the-quantum-measurement-problem-beb36fa099f9.

[8] N. Argaman, *A Lenient Causal Arrow of Time?*, Entropy, 20 (2018), p. 294.

[9] P. Ball, *Why the Many-Worlds Interpretation Has Many Problems*, Quanta Magazine, 18 October (2018).

[10] J. Bell, *On the Einstein–Podolsky–Rosen paradox*, Physics, 1 (1964), pp. 195–200.

[11] J. Bell, *Bertlmann's socks and the nature of reality*, Journal de Physique, C2 (1981), pp. 41–61.

[12] J. Bell, *Introductory Remarks*, Physics Reports, 137 (1986), pp. 7–9.

[13] M. Born, A. Einstein, and I. Born, *The Born Einstein Letters: Correspondence between Albert Einstein and Max and Hedwig Born from 1916 to 1955 with Commentaries by Max Born*, Macmillan Press, London, 1971. Translated by Irene Born.

[14] J. Bub and I. Pitowski, *Two dogmas about quantum mechanics*, in Many Worlds? Everett, Quantum Theory, and Reality, S. Saunders, J. Barrett, A. Kent, and D. Wallace, eds., Oxford University Press, Oxford, 2010, pp. 431–456.

[15] S. Carroll, *Mindscape 241: Tim Maudlin on Locality, Hidden Variables, and Quantum Foundations*, 2023. https://www.youtube.com/watch?v=8Y-JmocB84Y.

[16] J. Cramer, *The transactional interpretation of quantum mechanics*, Reviews of Modern Physics, 58 (1986), pp. 647–687.

[17] J. Cramer, *The Transactional Interpretation of Quantum Mechanics and Quantum Nonlocality*, in The Stanford Encyclopedia of Philosophy, E. N. Zalta, ed., Stanford University, 2015.

[18] P. C. W. Davies and J. R. Brown, *The Ghost in the Atom*, Cambridge University Press, Cambridge, 1986.

[19] B. S. DeWitt, *Quantum Mechanics and Reality*, Physics Today, 23 (1970), p. 30.

[20] H. Dukas and B. Hoffman, *Albert Einstein: The Human Side*, Princeton University Press, Princeton, NJ, 1979.

[21] D. Falk, *New Support for Alternative Quantum View*, Quanta Magazine, 16 May (2016).

[22] S. Friederich and P. W. Evans, *Retrocausality in quantum mechanics*, in Stanford Encyclopedia of Philosophy, E. N. Zalta, ed., Stanford University, 2019.

[23] C. Fuchs, *Quantum Mechanics as Quantum Information (and only a little more)*, 2002. Preprint. https://arxiv.org/abs/quant-ph/0205039.

[24] C. A. Fuchs, N. D. Mermin, and R. Schack, *An introduction to QBism with an application to the locality of quantum mechanics*, American Journal of Physics, 82 (2014), pp. 749–754.

[25] P. Goyal. *Derivation of Classical Mechanics in an Energetic Framework via Conservation and Relativity*, Foundations of Physics, 50 (2020), pp. 1426–1479. https://philarchive.org/archive/GOYDOC.

[26] J. Hance, S. Hossenfelder, and T. Palmer, *Supermeasured: Violating Bell-statistical Independence Without Violating Physical Statistical Independence*, Foundations of Physics, 52 (2022), p. 81.

[27] P. Höhn, *Toolbox for reconstructing quantum theory from rules on information acquisition*, 2018. Preprint. https://arxiv.org/abs/1412.8323.

[28] S. Hossenfelder, *Does Superdeterminism save Quantum Mechanics?*, 2021. https://www.youtube.com/watch?v=ytyjgIyegDI.

[29] R. Kastner, *Is there really "retrocausation" in time-symmetric approaches to quantum mechanics?*, AIP Conference Proceedings, 1841 (2017), p. 020002.

[30] D. Maltrana, M. Herrera, and F. Benitez, *Einstein's Theory of Theories and Mechanicism*, International Studies in the Philosophy of Science, 35 (2022), pp. 153–170.

[31] T. Maudlin, *Quantum Non-Locality and Relativity*, Wiley-Blackwell, Oxford, 2011.

[32] N. D. Mermin, *Making better sense of quantum mechanics*, Reports on Progress in Physics, 82 (2019), p. 012002.

[33] N. D. Mermin, *There is no quantum measurement problem*, Physics Today, 75 (2022), p. 62.

[34] NOVA Season 46 Episode 2, *Einstein's Quantum Riddle*, 2019. https://www.pbs.org/video/einsteins-quantum-riddle-ykvwhm/.

[35] J. Pienaar, *Comment on "The notion of locality in relational quantum mechanics"*, 2018. Preprint. https://arxiv.org/abs/1807.06457.

[36] S. Popescu and D. Rohrlich, *Quantum nonlocality as an axiom*, Foundations of Physics, 24 (1994), pp. 379–385.

[37] H. Price and K. Wharton, *Dispelling the Quantum Spooks – A Clue that Einstein Missed?*, 2013. Preprint. https://arxiv.org/abs/1307.7744.

[38] T. Ryckman, *Einstein*, Routledge, New York, 2017.

[39] L. Vaidman, *The Bell Inequality and the Many-Worlds Interpretation*, in Quantum Nonlocality and Reality: 50 Years of Bell's Theorem, M. Bell and S. Gao, eds., Cambridge University Press, Cambridge, 2016, pp. 195–203.

[40] L. Vaidman, *Many-Worlds Interpretation of Quantum Mechanics*, in The Stanford Encyclopedia of Philosophy, E. N. Zalta, ed., Stanford University, 2021.

[41] L. Vaidman, *Why the Many-Worlds Interpretation?*, 2022. Preprint. https://arxiv. org/abs/2208.04618.

[42] K. Wharton and N. Argaman, *Bell's theorem and locally mediated reformulations of quantum mechanics*, Reviews of Modern Physics, 92 (2020), p. 021002.

[43] K. Wharton and R. Liu, *Entanglement and the Path Integral*, 2022. Preprint. https://arxiv.org/abs/2206.02945.

[44] A. Zeilinger, *Quantum Information and Entanglement*, 2014. https://www.youtube. com/watch?v=TrsnMNh9W9U.

4

A Precedent: Special Relativity

The introduction of a "light ether" will prove superfluous, insofar as in accordance with the view to be developed here, no "space at absolute rest" endowed with special properties will be introduced.

Einstein (1905)

In this statement at the outset of his 1905 paper on special relativity [2], Einstein lets the reader know that he is going to solve the biggest mystery in physics of his time in a very radical fashion. That is, he is going to abandon the use of dynamical laws or mechanistic causal processes like the luminiferous aether to explain length contraction. Up to that point, physicists had only attempted constructive explanations of length contraction, none of which had won consensus support. This is analogous to what we saw in the last chapter with constructive solutions to the mystery of entanglement.

So, in this chapter we look at the mystery of length contraction and how Einstein solved it. What we will see is a possible way to solve the greatest mystery in physics today that avoids the pitfalls of constructive explanation. We will also see how the objective spacetime model of reality in special relativity opens the door for an all-at-once / atemporal / AGC explanation. While special relativity is where Einstein derived his famous equation $E = mc^2$, energy E equals mass m times c squared, we have no need of that because the mystery of length contraction deals only with kinematics (displacement, velocity, acceleration) not dynamics (kinematics + mass → momentum, force, energy, etc.). That mystery starts in 1865.

4.1 Maxwell, Michelson, and Morley

In 1865, James Clerk Maxwell published "A Dynamical Theory of the Electromagnetic Field," showing how electric fields can create magnetic fields and vice versa [8]. The equations he developed are called *Maxwell's equations* and they predicted the existence of electromagnetic waves propagating at a speed

Einstein's Entanglement. W. M. Stuckey, Michael Silberstein, and Timothy McDevitt, Oxford University Press.
© Oxford University Press (2024). DOI: 10.1093/9780198919698.003.0005

of about 300 000 km/s (or c for short). Waves usually imply the existence of a vibrating medium, e.g., sound waves are oscillating air and ocean waves are undulating water. So, physicists assumed the existence of a medium called the *luminiferous aether* to support electromagnetic waves.

It was very quickly apparent that the aether must have very unique (and problematic) properties. For example, the stiffer the medium, the faster waves will propagate through it and since c is very large, the aether must be millions of times stiffer than steel. However, as the planets orbit the Sun their orbits do not appear to be slowing down due to friction with the aether, so the aether must have negligible viscosity. For yet other reasons, physicists realized the aether also had to be massless, incompressible, and completely transparent. As the list of required properties grew, the aether seemed more and more like magic. If this was not enough to discard the idea of the aether, measurements of Earth's speed through the aether were producing an answer of zero. Indeed, all attempts to measure the speed of light were giving the same value c, regardless of one's motion relative to the source. This result violates Galilean (common-sense) velocity transformation between inertial reference frames.

For example, if you and I are at rest with respect to each other and I throw a rock away from myself in your direction at 15 m/s relative to me, then that rock is also moving towards you at 15 m/s. Now suppose I ride a bike towards you at 10 m/s and then throw the rock towards you at 15 m/s relative to me. The rock is now moving 15 m/s faster than the bike, which is already moving towards you at 10 m/s, so the speed of the rock relative to you is 25 m/s. That is Galilean velocity transformation from basic physics and it makes sense, right?

According to Galilean velocity transformation, if instead of throwing a rock at you I shine a flashlight at you while riding the bike at 10 m/s in your direction, then the light beam should be moving away from me at c and moving towards you at $c + 10$ m/s. But, what physicists were finding is that we both still measure the speed of that same light beam to be c! Now that could have been due to c being very large while everyday speeds (like the bike) are very small, so the difference between c and $c + 10$ m/s is just not detectable. While experimental physicists were not riding bikes, the speeds they were using, e.g., Earth's orbital speed about the Sun, were still much smaller than c. Enter two very skilled experimentalists, Albert Michelson and Edward Morley.

In 1887, Michelson and Morley devised a very precise way to detect Earth's motion through the aether using a device called an *interferometer* (Figure 4.1). The technical details do not concern us here, all you need to understand about this device is that the source's light beam is split into orthogonal paths (by a beam splitter), reflected by mirrors, and recombined (by the same beam splitter) to produce an interference pattern (a series of bright and dark strips or rings) that is viewed at the eyepiece. The interference pattern is caused by the light waves of the two beams being in or out of sync (phase) when they are recombined (Figure 4.2). Where the waves oscillate totally in phase you have a bright strip and where they oscillate totally out of phase you have a dark strip (Figure 4.2). Based on the idea of interference, the Michelson–Morley experiment is easy to understand.

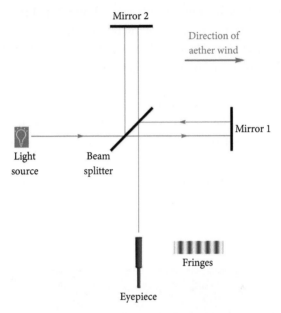

Fig. 4.1 Interferometer for the Michelson–Morley experiment.

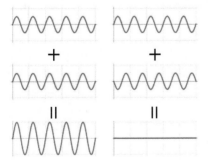

Fig. 4.2 Constructive interference results from waves being in sync (left side of picture). This produces the bright strips (or rings) of the interference pattern. Destructive interference results from waves being out of sync (right side of picture). This produces the dark strips (or rings) of the interference pattern.

Assuming the light moves in the aether at speed c and the interferometer is moving through the aether at speed v, then the light is moving relative to the interferometer at speeds between $c-v$ and $c+v$. For example, suppose the interferometer is moving in the aether horizontally as shown in Figure 4.1. Then the time the horizontal beam takes to go back and forth the horizontal distance L between the beam splitter and its mirror is $2L/c(1 - v^2/c^2)$. The time the vertical beam takes to go back and forth the vertical distance L' between the beam splitter and its mirror is $2L'/c\sqrt{1 - v^2/c^2}$. With $L = L'$ by construction, we see that the two beams take different times to go back and forth between the beam splitter and their respective mirrors and therefore arrive at the eyepiece out of phase. As the interferometer is rotated the two beams are affected differently; this changes their phase relationship when recombined, which changes the interference pattern.

Since light waves are very short and the interference pattern is based on the wavelength of light creating the pattern, this method can detect phase differences for relatively small v. The downside of the interferometer's extreme sensitivity was that phase changes could also be created by very small unrelated disturbances, like horse-drawn carriages on the street outside the building. To counter this, Michelson and Morley constructed their interferometer on a huge slab of stone floating in a pool of mercury (Figure 4.3).

Fig. 4.3 Photo of Michelson and Morley's interferometer. [Case Western Reserve University. Public domain, via Wikimedia Commons.]

This allowed their interferometer to unambiguously detect the amount of phase change expected by v as small as the Earth's orbital speed about the Sun.

Indeed, Michelson and Morley made L large enough (by multiple reflections inside the interferometer) that with v equal to Earth's orbital speed about the Sun they would observe a change of 0.4 fringes[1] when rotating the interferometer. With the incredible sensitivity of their interferometer they were able to measure a difference of 0.01 fringes, so 0.4 fringes was easy to detect even though the speed v giving rise to it was only $v/c \approx 10^{-4}$.

Given this brilliant experimental design, Michelson and Morley were shocked to find no change in the interference pattern as they rotated their interferometer. They repeated the experiment at different times of the day and on different days of the year to make sure they did not happen to do the experiment when v was zero, i.e., when Earth just happened to be at rest with respect to the aether. Nonetheless, their famous "null result" always obtained. That was the state of affairs in 1887.

4.2 "Constructive Efforts" to Explain Length Contraction

Just as physicists today have tried unsuccessfully to solve the mystery of entanglement constructively, physicists in the late 19th century tried unsuccessfully to solve the mystery of Michelson and Morley's null result constructively. Patrick Moylan writes [9]:

> To put things into an historical perspective, we recall that at the end of the nineteenth century, physics was in a terrible state of confusion. Maxwell's equations were not preserved under the Galilean transformations, and most of the Maxwellian physicists of the time were ready to abandon the relativity of motion principle ... They adopted a distinguished frame of reference, the rest frame of the "luminiferous aether," as the medium in which electromagnetic waves propagate and in which Maxwell's equations and the Lorentz force law have their usual forms. In effect, they were ready to uproot Copernicus and reinstate a new form of geocentrism.

For example, Oliver Heaviside showed that a charge's electric field would be distorted when the charge was moving in the aether. Since an object is made

---

[1] A fringe is the distance between two bright strips in the interference pattern.

of charged particles held together by their electric fields, George FitzGerald and Hendrik Lorentz conjectured that an object's length would shrink along the direction of its motion in the aether. If the horizontal length L in Figure 4.1 shrinks by $\sqrt{1 - v^2/c^2}$, i.e., $L \to L\sqrt{1 - v^2/c^2}$, while the vertical length $L' = L$ does not change at all, then the times for the two beams to go back and forth between the beam splitter and their respective mirrors are equal, explaining Michelson and Morley's null result.

Joseph Larmor and Lorentz then asked how Maxwell's equations would be affected by this FitzGerald–Lorentz (or length) contraction and discovered that in order for Maxwell's equations to be invariant under length contraction, one would also have to introduce *time dilation*. That is, clocks moving through the aether would run slower than those at rest with respect to the aether by the same factor, $\sqrt{1 - v^2/c^2}$.

Henri Poincaré pointed out that these *Lorentz transformations*, as he named them, form a mathematical structure called a *group*. This group contains transformations between reference frames in relative motion (Lorentz boosts) and transformations between reference frames rotated in space with respect to each other (rotations). In this group, the commutator of two rotations is a rotation and the commutator of two boosts is a rotation. So, rotations can stand alone as their own group, but Lorentz boosts need rotations to form a group. Poincaré added spatial and temporal translations to this list of transformations between inertial reference frames to form what is now known as the *Poincaré symmetry group*.

Below, we will see how the boost invariance of measurements of c give precisely the factor $\sqrt{1 - v^2/c^2}$ involved in length contraction and time dilation. While c must be invariant under all transformations between inertial reference frames, the boost invariance leads to the mystery of length contraction because c is a velocity and velocity generates boosts. To say a velocity must have the same value at different times or in different locations or in different directions is not so mysterious, but to say it must have the same value in reference frames that are themselves moving at different relative speeds violates our intuition à la Galilean velocity transformation, as described above.

Given that Maxwell's equations were invariant under Lorentz transformations and Lorentz transformations entailed the length contraction needed to explain the Michelson–Morley experiment, physicists naturally tried to find a physical justification for them. But, the ridiculously unverifiable physical properties of the aether kept physicists from buying into any particular version of aether theory. Sound familiar?

Einstein's attempt to account constructively for the Michelson–Morley null result involved emission theories of light. Rather than have light maintain a speed c with respect to the aether, emission theories held that light maintained its speed c relative to the emitter. Of course, this would easily explain the Michelson–Morley null result because their light source was attached to the interferometer. John Norton notes that even [10, p. 38]

> Einstein was willing to sacrifice the greatest success of 19th century physics, Maxwell's theory, seeking to replace it by one conforming to an emission theory of light, as the classical, Galilean kinematics demanded.

However, emission theories had their own problems, e.g., the light emitted from stars did not conform to emission theory predictions, so physicists were not sold on them either. Then with the successful Lorentz transformations as now with the successful theory of quantum mechanics, physicists became increasingly desperate to find a constructive explanation for an otherwise successful formalism. Norton writes [10, p. 38]:

> With the failures mounting and his options exhausted, Einstein would entertain an extraordinary and desperate thought. Could he realize the principle of relativity in electrodynamics if he reshaped the very notion of time itself?

This is what led Einstein to propose his radical solution to the mystery of length contraction that we now know as the theory of special relativity.

4.3 Einstein's Principle Solution

As we pointed out in the preface, Einstein wrote [3]:

> By and by I despaired of the possibility of discovering the true laws by means of constructive efforts based on known facts. The longer and the more despairingly I tried, the more I came to the conviction that only the discovery of a universal formal principle could lead us to assured results.

In other words, Einstein simply gave up trying to explain length contraction via dynamical laws or mechanistic causal processes ("constructive efforts"). Instead, he used a compelling fundamental principle, i.e., the relativity principle, to solve the mystery of length contraction. As Paul Mainwood writes [6], most people do not even understand that special relativity is a principle theory, i.e., "there is no mention in relativity of exactly *how* clocks slow, or *why* meter sticks shrink" (no "constructive efforts"). Nonetheless, the relativity principle is so compelling that physicists have all but stopped working on a constructive counterpart to special relativity. As Maudlin writes [7, p. 221]:

> For example, it is possible to design theories that are empirically equivalent to the Special Theory of Relativity but that posit Newtonian Absolute Space and Absolute Time. If one supposes that Maxwell's equations hold in only the One True Reference Frame one can then derive that the behavior of electromagnetic clocks and measuring rods will not allow one to discover which inertial reference frame is the One True One. Rods will shrink and clocks will slow down in just such a way that the speed of light *seems* to be the same in all frames, though it is not. Such a theory, although logically consistent and empirically impeccable, is generally considered to be inferior to Special Relativity. The grounds for this judgement are not usually made very explicit, but the general idea is that it would be awfully deceptive to create a world with Absolute Space and then use the laws of physics to hide its existence from us.

So, it is no great surprise that this principle understanding of special relativity is presented in introductory physics textbooks [5, 13]. Those authors note that Einstein's relativity principle, "the laws of physics must be the same in all inertial reference frames," is generalized from Galileo's relativity principle, "The laws of mechanics must be the same in all inertial reference frames." They then argue that the relativity principle justifies the *light postulate*:

> Everyone must measure the same speed of light c, regardless of their motion relative to the source.

If there were only one reference frame for a source in which the speed of light equalled the prediction from Maxwell's equations, then that would certainly constitute a preferred reference frame. The mystery of length contraction is then easy to explain as follows.

Suppose Alice is riding in a wagon at speed v relative to Bob (Figure 4.4). Alice turns on a light source a distance D below a mirror on the wagon so that the light beam moves vertically upwards a distance D, reflects off the mirror, then travels downwards a distance D, returning to the source in time ΔT, as measured by Alice. While the light beam travels only vertically in Alice's reference frame, it has a horizontal component in Bob's reference frame given

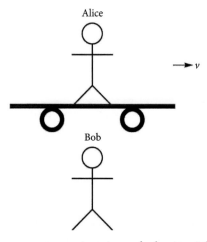

Fig. 4.4 Alice is on a wagon moving at speed v horizontally relative to Bob.

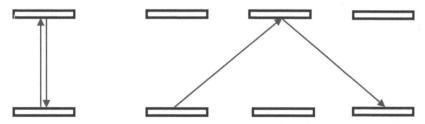

Fig. 4.5 The light beam travels only vertically in Alice's reference frame (left side). In Bob's reference frame, the light beam also has a horizontal component, so the light beam has a diagonal path in space in Bob's reference frame (right side).

by $v\Delta t$ (Figure 4.5), where Δt is the time Bob measures between the emission event and reception event for the light beam.

So, the distance the light travels from the emission event to the reception event is $2\sqrt{D^2 + v^2\Delta t^2/4}$ in Bob's reference frame (twice the hypotenuse length per the Pythagorean theorem, Figure 4.6). Since Alice and Bob must both measure the same speed c for the light beam, Alice has $c = 2D/\Delta T$ while Bob has $c = (2\sqrt{D^2 + v^2\Delta t^2/4})/\Delta t$. Solving for ΔT, we find the time Alice measures between the emission event and the reception event ΔT is shorter than Δt that Bob measures, i.e., $\Delta T = \sqrt{1 - v^2/c^2}\Delta t$.

Of course, you do not need all the equations to see that Alice's time between the events must be shorter than Bob's. All you have to do is look at Figure 4.5 and see that the distance between the events is shorter for Alice than for Bob. Since Alice's shorter distance divided by Alice's time must equal Bob's longer

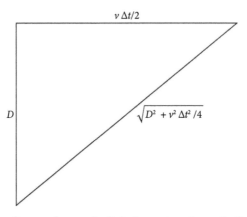

Fig. 4.6 In Bob's reference frame, the light beam travels vertically a distance D and horizontally a distance $v\Delta t/2$ in traveling from the source to the mirror. So, the light beam has a diagonal path of $\sqrt{D^2 + v^2\Delta t^2/4}$ in Bob's reference frame.

distance divided by Bob's time, i.e., both must measure the same speed for the light beam, then Alice's time must be shorter than Bob's time.

The two events (emission and reception of the light signal) occur at the same place in space for Alice, so ΔT is called the *proper time* between the events. Observers in reference frames that see Alice moving relative to them see the two events occur at different spatial locations and measure a longer time between the events, i.e., the proper time is "dilated." If Alice is using this bouncing light beam as a clock, those observers will say Alice's clock is running slow by a factor of $\sqrt{1 - v^2/c^2}$. That is why you often hear people say moving clocks run slow according to special relativity per time dilation, and we can use this time dilation to derive length contraction.

To do that, suppose the front of Alice's wagon passes Bob at Event A at $t = T = 0$ (Figure 4.7) and the back of Alice's wagon passes Bob at Event B at time t (Figure 4.8). Bob says the wagon is moving at a speed of ℓ/t where ℓ is the length of the wagon according to Bob. Since Events A and B occurred at the same spatial location for Bob, t is the proper time between the events and will be shorter than the time Alice measures between the events by the factor $\sqrt{1 - v^2/c^2}$, i.e., $t = \sqrt{1 - v^2/c^2}\,T$. If we want Bob's computation of Alice's speed relative to him to equal Alice's computation of Bob's speed relative to her, then Bob's length for the wagon ℓ must be shorter than Alice's length of the wagon L by the exact same factor, i.e., $\ell = \sqrt{1 - v^2/c^2}\,L$.

Since the wagon is at rest with respect to Alice, her length L is called the *proper length* and everyone in motion with respect to her along the wagon

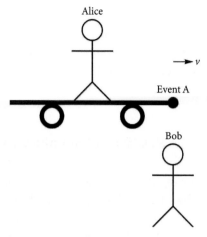

Fig. 4.7 The front of Alice's wagon passes Bob at $t = T = 0$ (Event A).

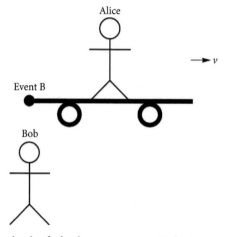

Fig. 4.8 The back of Alice's wagon passes Bob at time t (Event B).

will measure a shorter length for the wagon, i.e., the proper length is "contracted." That is why you often hear people say that moving meter sticks are shortened according to special relativity per length contraction. That is how length contraction follows from NPRF + c.

Note the difference between Lorentz's constructive explanation of the Michelson–Morley null result and Einstein's principle explanation. Lorentz needs to invoke a medium with impossible physical characteristics that causes the horizontal arm of the interferometer to shrink by $\sqrt{1 - v^2/c^2}$. Demanding that Maxwell's equations be invariant under length contraction then leads to time dilation. Conversely, Einstein invokes the relativity principle which directly dictates the Michelson–Morley null result, i.e., everyone must measure the same value of c, regardless of their motion relative to the source. This light postulate then leads to time dilation and length contraction as shown above. Einstein's principle account flips the explanatory sequence of Lorentz's constructive account.

4.4 The Relativity of Simultaneity

But, wait a minute. If moving clocks run slow, then Bob and observers at rest with respect to him (Bob's reference frame) will say clocks in Alice's reference frame are running slow, while observers in Alice's reference frame will have to say the same thing about clocks in Bob's reference frame. How can the clocks in Alice's reference frame run slower than those in Bob's reference frame while the converse is also true?

Likewise, observers in Bob's reference frame will have to say meter sticks in Alice's reference frame are short, while observers in Alice's reference frame will have to say the same thing about meter sticks in Bob's reference frame. How can the meter sticks in Alice's reference frame be shorter than those in Bob's reference frame while the converse is also true? The resolution of this paradox resides in the relativity of simultaneity, as we introduced in Chapter 1.

Consider the following four events adapted from the 2004 version by Richard DeWitt [1] (who adapted his version from Mermin) using exaggerated time differences:

- Event 1: 20 year-old Joe and 20 year-old Sara meet.
- Event 2: 20 year-old Bob and 17.5 year-old Alice meet.
- Event 3: 22 year-old Bob and 20 year-old Kim meet.
- Event 4: 25.6 year-old Bob and 24.5 year-old Sara meet.

The girls and the boys agree on the facts contained in these four events. Further, Joe and Bob see the girls moving in the positive x direction (Figure 4.9), so the girls see the boys moving in the negative X direction at the same speed (Figure 4.10). Additionally, the boys are the same age in their reference frame and the girls are the same age in their reference frame. This establishes simultaneity for each set, i.e., events are simultaneous (coexist) for the boys if the events occur when the boys are the same age, e.g., Events 1 and 2 above. Likewise for the girls, e.g., Events 1 and 3 above.

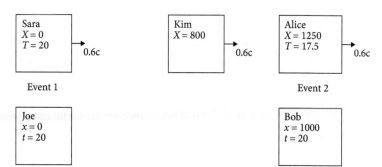

Fig. 4.9 The girls are the same age in their reference frame and the boys are the same age in their reference frame. In the figures that follow, we exaggerate the time differences for effect. The temporal origin corresponds to being 20 years old and –0.0025 s corresponds to being 2.5 years younger than 20 years old, i.e., 17.5 years old, and so on. The distances are in kilometers. This figure shows Events 1 and 2 occurring simultaneously from the boys' perspective (see Figure 4.13).

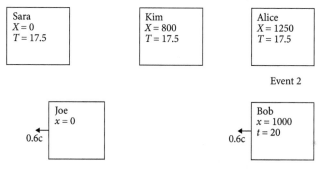

Fig. 4.10 The girls do not believe Events 1 and 2 are simultaneous, i.e., Event 2 occurred 2.5 years before Event 1 according to the girls (see Figure 4.14).

In fact, Einstein specifically defined time using simultaneity [2]:

> We have to take into account that all our judgments in which time plays a part are always judgments of *simultaneous events*. [italics in original]

His notion of simultaneity was that of the synchronicity of *stationary* clocks, which was established by exchanging light signals [2]:

> Thus with the help of certain imaginary physical experiments [associated with the exchange of light signals] we have settled what is to be understood by synchronous stationary clocks located at different places, and have evidently obtained a definition of "simultaneous," or "synchronous," and of

"time." The "time" of an event is that which is given simultaneously with the event by a stationary clock located at the place of the event, this clock being synchronous, and indeed synchronous for all time determinations, with a specified stationary clock. ...

It is essential to have time defined by means of stationary clocks in the stationary system, and the time now defined being appropriate to the stationary system we call it "the time of the stationary system."

So, the boys' clocks (mechanical and biological) are stationary and synchronized with respect to each other and establish "the time of [their] stationary system," while the same is true for the girls' clocks (mechanical and biological). Accordingly, the boys are the same age in their stationary system and say they coexist when they are the same age, while the girls are the same age in their stationary system and would make the same claim about their coexistence.

Again, as explained in Chapter 1, the set of coexisting bodily objects and events in spacetime is called a NOW slice and observers in reference frames moving relative to each other do not agree as to how to carve Minkowski spacetime into NOW slices [4]. For example, Events 1 and 2 reside on a NOW slice for the boys, but not for the girls. Likewise, Events 1 and 3 reside on a NOW slice for the girls, but not for the boys. Let's look at the implications of this relativity of simultaneity for the four events above.

We start with the boys' perspective for Events 1 and 2 (Figure 4.9). Since the boys say Events 1 and 2 are simultaneous (reside on the same NOW slice), they must conclude that the girls are not the same age, i.e., Alice is 2.5 years younger than Sara. Second, they conclude that the girls' meter sticks are short because the girls say the distance between Sara and Alice is 1250 km while the boys say it is only 1000 km.

Of course, the girls disagree—they have celebrated birthdays together for many years and they know for a fact that they are the same age. Event 2 is definitely not simultaneous with Event 1, it happened 2.5 years before Event 1 (Figure 4.10). The event simultaneous with Event 1 is Event 3 as far as the girls are concerned, i.e., Events 1 and 3 reside on the same NOW slice for the girls (Figure 4.11). So, they conclude that the boys are not the same age, i.e., Bob is two years older than Joe. Second, they conclude that the boys' meter sticks are short because the boys say the distance between Joe and Bob is 1000 km while the girls say it is only 800 km. This is how the relativity of simultaneity allows for the boys to claim the girls' meter sticks are short and the girls to claim the boys' meter sticks are short. So, we see how the relativity of simultaneity resolves the paradox concerning length contraction. What about time dilation?

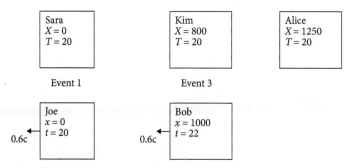

Fig. 4.11 The girls say Events 1 and 3 are simultaneous and the boys are not the same age (see Figure 4.14).

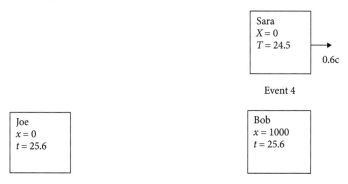

Fig. 4.12 Event 4 according to the boys (see Figure 4.13).

The girls see clearly that the boys' clocks are running slow because the girls aged 2.5 years between the time Bob and Alice met (Figure 4.9) and the time Bob and Kim met (Figure 4.11), while Bob only aged 2 years. Again, this conclusion is based on the fact that the girls are the same age, which the boys obviously deny. Rather, the boys say it is the girls who are aging more slowly because the boys aged 5.6 years between the time Joe and Sara met (Figure 4.9) and the time Bob and Sara met (Figure 4.12), while Sara only aged 4.5 years. Again, this conclusion is based on the fact that the boys are the same age, which the girls obviously deny.

4.5 Minkowski Spacetime

In conclusion, the boys say the girls' clocks run slow and the girls' meter sticks are short, while the girls say exactly the same thing about the boys' clocks and meter sticks. These paradoxical facts can both be true because the boys and girls disagree as to which events are simultaneous—Events 1 and 2 are simultaneous (reside on the same NOW slice, i.e., coexist) for the boys (Figure 4.9) while Events 1 and 3 are simultaneous (reside on the same NOW slice, i.e., coexist) for the girls (Figure 4.11). Consequently, the boys believe they are the same age and deny that the girls are the same age, and vice versa.

We saw this same relativity of perspective between Bob and Alice when they occupied different inertial reference frames (differing in spatial orientation) for the quantum gloves experiment in Chapter 0. There, Bob told Alice that she had to average her spin measurement outcomes to satisfy conservation of spin, while Alice told Bob that *he* had to average *his* spin measurement outcomes to

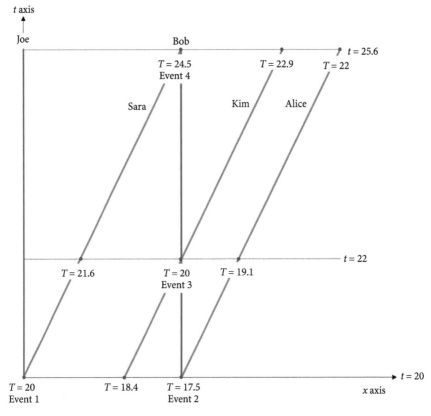

Fig. 4.13 The boys' (3+1)D spacetime model showing their worldlines (blue) and the girls' worldlines (brown). According to the boys, Kim is 1.6 years younger than Sara and Alice is 2.5 years younger than Sara. That means the girls are all different ages on any of the boys' NOW slices as determined by the boys' proper (and coordinate) time *t*. All the girls age 80% slower than the boys. The *t* = 20, 22, and 25.6 NOW slices constitute coexisting bodily objects and events according to the boys.

satisfy conservation of spin. We will revisit that for actual spin measurements in Chapter 7.

Of course, if the boys' and girls' perspectives are equally valid in accord with NPRF, then we must reject the absolute space and absolute time in Newton's objective spacetime model of reality. Again, the game of physics deals fundamentally with observers who collect, exchange, and coherently synthesize the information from their individual data collection devices into an objective (intersubjectively-agreed-upon) spacetime model of reality. Accordingly, since observers in Bob's reference frame disagree with observers in Alice's reference frame about what constitutes coexisting bodily objects and events (NOW slice of spacetime, Figures 4.13 and 4.14) per the relativity of simultaneity, our new

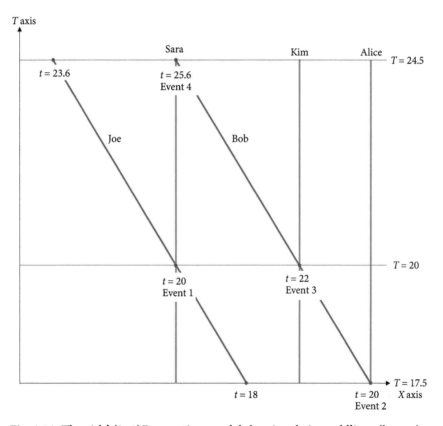

Fig. 4.14 The girls' (3+1)D spacetime model showing their worldlines (brown) and the boys' worldlines (blue). According to the girls, Joe is two years younger than Bob. That means the boys are different ages on any of the girls' NOW slices as determined by the girls' proper (and coordinate) time T. The boys age 80% slower than the girls. The $T = 17.5$, 20, and 24.5 NOW slices constitute coexisting bodily objects and events according to the girls.

objective spacetime model of reality must allow for relative notions of space and time. And, in 1907 Hermann Minkowski did just that.

Specifically, Minkowski showed how to fuse all of the individual subjective three-dimensional space plus one-dimensional time ((3+1)D) spacetimes into a single four-dimensional (4D) spacetime now called *Minkowski space*, *Minkowski spacetime*, or *M4* (Figure 4.15). This objective spacetime model of reality provides a single, self-consistent 4D spacetime model of all the

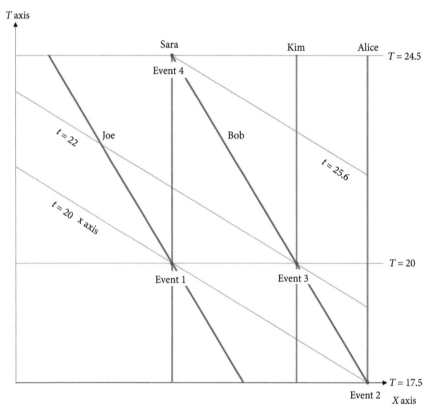

Fig. 4.15 Minkowski spacetime for Events 1, 2, 3, and 4 showing the boys' and girls' paths in spacetime (their worldlines). Notice the boys' coordinate time *t* gives their age (proper time) while the girls' coordinate time *T* gives their age (proper time). The girls' NOW slices for *T* = 17.5, 20, and 24.5 and the boys' NOW slices for *t* = 20, 22, and 25.6 are shown. In this version of M4, the girls' worldlines are vertical while the boys' are slanted. In other versions related by Lorentz transformations, both sets of worldlines could be slanted or the boys' worldlines could be vertical and girls' slanted, as in Figure 4.13.

events from all the different reference frames of all the different data collection devices. That is, this objective 4D spacetime model of reality reconciles all of the individual subjective (3+1)D spacetime models, even though each individual (3+1)D spacetime model organizes (partitions) the events into different past, present, and future spacetime relationships (relativity of simultaneity) as necessitated by NPRF (Figures 4.13 and 4.14).

4.6 Summary

The greatest mystery in physics at the end of the 19th century was length contraction, which was apparently needed to explain why everyone was measuring the same value for the speed of light c, regardless of their relative motions (light postulate). The mystery existed precisely because everyone believed that some constructive mechanism (like the aether) was needed to 'cause' the light postulate. That is, a constructive explanation was needed such as

aether → length contraction → light postulate.

Einstein produced an entirely different type of solution to the mystery of length contraction by turning to principle/structural explanation:

relativity principle → light postulate → length contraction.

His principle solution (theory of special relativity) to the greatest mystery in physics of his day is universally accepted by physicists even though we still have no (consensus) constructive counterpart.

Another key aspect of special relativity is its challenge to our everyday thinking about reality. That is, special relativity replaces the absolute simultaneity of coexisting bodily objects and events in Newton's objective spacetime model of reality with the relative simultaneity of coexisting bodily objects and events in Minkowski's objective spacetime model of reality [11]. These two aspects of special relativity open the door to a principle, AGC solution to the greatest mystery in physics today, i.e., entanglement. All of this means there is an upper limit $c \neq \infty$ to the speed of information transfer.

While it is true that the spacetime of special relativity preserves the temporal order of causally (timelike) related events between different inertial reference frames, we are left to wonder, should constructive explanation still be considered fundamental to principle explanation? We answer that question in the next chapter.

But before leaving this chapter, we want to share an interesting historical fact provided by Thomas Ryckman [12]:

> Despite the relativity (or frame dependence) of measures of lengths and times (indeed, all physical quantities with the dimensions of lengths or times), the invariance of the interval shows that relativity theory is really misnamed. Not by Einstein. He had proposed the term invariant theory before learning that this term had already been appropriated within pure mathematics. The designation relativity theory is due to Max Planck, who sought to emphasize the universal scope of the principle of relativity. Planck's term caught on, and very quickly it was too late for any rebaptism.

That Planck was responsible for the name "relativity theory" because he was keen to emphasize "the universal scope of the principle of relativity" turns out to be quite ironic. As we will see in Chapters 6 and 7, applying the relativity principle to the measurement of his constant h will solve the mystery of entanglement and thereby answer Mermin's challenge and resolve the EPR and EPR–Bell paradoxes. Here, $h \neq 0$ restricts the amount of simultaneous information available for a quantum.

References

[1] R. DeWitt, *Worldviews: An Introduction to the History and Philosophy of Science*, Blackwell Publishing, Oxford, 2004.

[2] A. Einstein, *On the Electrodynamics of Moving Bodies*, Annalen der Physik, 17 (1905), pp. 891–921.

[3] A. Einstein, *Autobiographical notes*, in Albert Einstein: Philosopher-Scientist, P. Schilpp, ed., Open Court, La Salle, IL, 1949, pp. 3–94.

[4] B. Greene, *The Fabric of the Cosmos: Space, Time, and the Texture of Reality*, Vintage, New York, 2004.

[5] R. Knight, *Physics for Scientists and Engineers with Modern Physics*, Pearson, San Francisco, CA, 2022.

[6] P. Mainwood, *What Do Most People Misunderstand About Einstein's Theory Of Relativity?*, 2018. https://www.forbes.com/sites/quora/2018/09/19/what-do-most-people-misunderstand-about-einsteins-theory-of-relativity.

[7] T. Maudlin, *Quantum Non-Locality and Relativity*, Wiley-Blackwell, Oxford, 2011.

[8] J. C. Maxwell, *A Dynamical Theory of the Electromagnetic Field*, Philosophical Transactions of the Royal Society of London, 155 (1865), pp. 459–512.

[9] P. Moylan, *Velocity reciprocity and the relativity principle*, American Journal of Physics, 90 (2022), pp. 126–134.

[10] J. Norton, *Einstein's Special Theory of Relativity and the Problems in the Electrody-
      namics of Moving Bodies that Led Him to It*, in The Cambridge Companion to Ein-
      stein, M. Janssen and C. Lehner, eds., Cambridge University Press, Cambridge, 2014,
      pp. 72–102.

[11] D. Peterson and M. Silberstein, *Relativity of Simultaneity and Eternalism: In Defense
      of the Block Universe*, in Space, Time, and Spacetime – Physical and Philosophical
      Implications of Minkowski's Unification of Space and Time, V. Petkov, ed., Springer,
      Berlin, 2010, pp. 209–237.

[12] T. Ryckman, *Einstein*, Routledge, New York, 2017.

[13] R. Serway and J. Jewett, *Physics for Scientists and Engineers*, Brooks/Cole, Boston,
      MA, 10th ed., 2019.

5

Einstein's Principle or Reichenbach's?

> If A and B are correlated, then A is a cause of B; B is a cause of A; or A
> and B are both caused by a third factor, C. In the last case, the common
> cause C occurs prior to A and B.
>
> Reichenbach's Common Cause Principle
> (Stanford Encyclopedia of Philosophy, 2020)

In Chapters 0–2, we showed you the mystery of entanglement and the Bell inequality according to Albert's quantum gloves experiment and the Mermin device using the quantum property of spin angular momentum. We also introduced you to constructive/mechanistic versus principle and structural explanation, the EPR paradox, the EPR–Bell paradox, and Mermin's challenge.

In Chapter 3, we introduced the (quite reasonable) constructive bias via Newtonian mechanics by contrasting explanation using dynamical laws with explanation using conservation principles. We also outlined the main constructive efforts to solve the mystery in order to resolve the paradoxes and answer the challenge. Hopefully, it was clear at that point why the mystery of entanglement remains unsolved after being introduced by Einstein almost 90 years ago. Perhaps you can even sympathize with those who call it "the greatest mystery in physics."

The bottom line is that the foundations community has reached an impasse in trying to produce a consensus solution to the mystery of entanglement because of conflicting expectations. On the one hand the foundations community wants a constructive solution to the mystery in accord with Reichenbach's Principle, while on the other hand a consensus solution can only be achieved if it does not violate locality, statistical independence, intersubjective agreement, or unique experimental outcomes. Unfortunately, Bell's theorem and the experimental violation of Bell's inequality tell us that any constructive solution to the mystery of entanglement will necessarily violate at least one of those four characteristics. Therefore, to resolve the impasse we will need to give up at least one of these two expectations.

Einstein's Entanglement. W. M. Stuckey, Michael Silberstein, and Timothy McDevitt, Oxford University Press.
© Oxford University Press (2024). DOI: 10.1093/9780198919698.003.0006

This is an ironic state of affairs because as Bertrand Russell [33, pp. 171–196] and many others such as Huw Price [28] have argued, there is nothing overtly causal about dynamical explanations via differential equations, and dynamical explanations in physics are complete in terms of explaining why new events come into being. That is why many physicists and philosophers believe causal explanations are strictly perspectival and pragmatic: they only make sense from the point of view of causal agents attempting to intervene in the world, such as choosing measurement settings. But again, as we discuss at length elsewhere [36, 37], even those with this point of view are loath to completely abandon constructive explanation in all its causal and dynamical forms in exchange for fully all-at-once AGC-type explanations.

In this chapter, we will argue for giving up the constructive bias. We understand that this bias is not easy to relinquish, e.g., as we will see in the next section, even Einstein shared this bias. However, given that NPRF + c is a consensus principle solution to the mystery of length contraction (Chapter 4), and given that the principle solution to the mystery of entanglement that we are offering is NPRF + h, which does not violate locality, statistical independence, intersubjective agreement, or unique experimental outcomes, we believe it is reasonable to give up Reichenbach's Principle and adopt Einstein's relativity principle for solving the greatest mystery in physics.

5.1 Constructive Bias

Widespread acceptance of Einstein's 1905 principle account of length contraction was many years in the making. For example, questions on the Mathematical Tripos examination at Cambridge University contained reference to various "jelly, froth, and vortex" models of the aether until 1909 [21, pp. 236–240]. The following 1906 quote from Lorentz sums up nicely the initial response to Einstein's principle solution, foreshadowing the eventual acceptance of its beauty [24, p. 230]:

> It will be clear by what has been said that the impressions received by the two observers A0 and A would be alike in all respects. It would be impossible to decide which of them moves or stands still with respect to the ether, and there would be no reason for preferring the times and lengths measured by the one to those determined by the other, nor for saying that either of them is in possession of the "true" times or the "true" lengths. This is a point which Einstein has laid particular stress on, in a theory in which he starts from what he calls the principle of relativity ...

I cannot speak here of the many highly interesting applications which Einstein has made of this principle. His results concerning electromagnetic and optical phenomena agree in the main with those which we have obtained in the preceding pages, the chief difference being that Einstein simply postulates what we have deduced, with some difficulty and not altogether satisfactorily, from the fundamental equations of the electromagnetic field. By doing so, he may certainly take credit for making us see in the negative result of experiments like those of Michelson, Rayleigh and Brace, not a fortuitous compensation of opposing effects, but the manifestation of a general and fundamental principle.

While Lorentz complains that Einstein "simply postulates what we have deduced," he ends by acknowledging that Einstein made "us see ... the manifestation of a general and fundamental principle." As we pointed out in Chapter 1, the explanatory hierarchy in constructive explanation is flipped in the corresponding principle explanation. For the mystery of length contraction, the constructive explanation is

$$\text{aether } \rightarrow \text{ length contraction } \rightarrow \text{ light postulate,}$$

while its principle counterpart is

$$\text{relativity principle } \rightarrow \text{ light postulate } \rightarrow \text{ length contraction.}$$

Obviously, the key to the success of a principle explanation resides in the compelling nature of its fundamental principle and, as we explained in Chapter 1, physicists find the relativity principle to be compelling.

That is why today, nearly 120 years after Einstein published special relativity, physicists are content with its principle explanation of length contraction and very few physicists are looking for a constructive counterpart. Indeed, Einstein's principle solution is so widely accepted that introductory physics textbooks introduce special relativity in principle fashion without complaining about the absence of causal mechanisms such as the aether or emission theories of light [23, 35]. Therefore, it may surprise you to know that Einstein himself believed special relativity needed a constructive counterpart.

Here is what Einstein said to Arnold Sommerfeld in 1908 [18]:

It seems to me too that a physical theory can be satisfactory only when it builds up its structures from *elementary* foundations. The theory of relativity is not more conclusively and absolutely satisfactory than, for example, classical thermodynamics was before Boltzmann had interpreted entropy as probability. If the Michelson–Morley experiment had not put us in the worst predicament, no one would have perceived the relativity theory as a (half) salvation. Besides, I believe that we are still far from satisfactory elementary foundations for electrical and mechanical processes. I have come to this pessimistic view mainly as a result of endless, vain efforts to interpret the second universal constant [h] in Planck's radiation law in an intuitive way.

As Harvey Brown and Christopher Timpson point out [8], "Einstein was emphasizing the *limitations* of [special relativity], not its strengths" in his statement. Brown and Timpson produce many more statements by Einstein along these lines, so Einstein clearly believed special relativity needed a constructive completion. Here is how he described constructive versus principle theories [16]:

We can distinguish various kinds of theories in physics. Most of them are constructive. They attempt to build up a picture of the more complex phenomena out of the materials of a relatively simple formal scheme from which they start out. Thus the kinetic theory of gases seeks to reduce mechanical, thermal, and diffusional processes to movements of molecules—i.e., to build them up out of the hypothesis of molecular motion. When we say that we have succeeded in understanding a group of natural processes, we invariably mean that a constructive theory has been found which covers the processes in question.

Along with this most important class of theories there exists a second, which I will call "principle-theories." These employ the analytic, not the synthetic, method. The elements which form their basis and starting point are not hypothetically constructed but empirically discovered ones, general characteristics of natural processes, principles that give rise to mathematically formulated criteria which the separate processes or the theoretical representations of them have to satisfy. Thus the science of thermodynamics seeks by analytical means to deduce necessary conditions, which separate events have to satisfy, from the universally experienced fact that perpetual motion is impossible.

The advantages of the constructive theory are completeness, adaptability, and clearness, those of the principle theory are logical perfection and security of the foundations. The theory of relativity belongs to the latter class.

In particular, note that he writes, "When we say that we have succeeded in understanding a group of natural processes, we invariably mean that a constructive theory has been found which covers the processes in question."

Most agree that Einstein believed constructive explanation is fundamental to principle explanation. For Einstein, the role of principle explanation in the absence of a constructive explanation is to place constraints on any forthcoming constructive explanation, thereby narrowing the constructive possibilities. For instance, Einstein hoped that Minkowski spacetime would merely constrain some future constructive explanation for time dilation and length contraction. Of course, no such constructive counterpart to special relativity has ever been adopted, nor is there any real effort being made to find one, but Einstein died without the benefit of knowing the status of special relativity today. He also died well before seeing how his relativity principle could be used with "the second universal constant [h] in Planck's radiation law" to resolve his EPR paradox.

One of Einstein's favorite examples of constructive theories versus principle theories was statistical mechanics (constructive) versus thermodynamics (principle). In his statement above, he points out that the empirically discovered fact at the foundation of thermodynamics is "the universally experienced fact that perpetual motion is impossible." He also had an example along those lines as to how a principle explanation constrained and aided the development of a corresponding constructive theory.

Regarding Boltzmann's principle from statistical mechanics (you do not need to know exactly what that is), which he mentioned in his letter to Sommerfeld above, Einstein wrote [22]:

[Boltzmann's principle] connects thermodynamics with the molecular theory. It yields, as well, the statistical probabilities of the states of systems for which we are not in a position to construct a molecular-theoretical model. To that extent, Boltzmann's magnificent idea is of significance for theoretical physics ... because it provides a heuristic principle whose range extends beyond the domain of validity of molecular mechanics.

In this case, blackbody radiation involving quanta of electromagnetic energy (introduced in Chapter 2), taken from the constructive quantum model of radiation, satisfies Boltzmann's principle. In other words, Boltzmann's principle was the constraint that initially suggested Planck and Einstein's constructive quantum model of radiation.

Along these lines, Brown and Timpson's paper [8] cited above is titled "Why special relativity should not be a template for a fundamental reformulation of quantum mechanics." They make their case using the now famous 2003 information-theoretic reconstruction of quantum mechanics by Rob Clifton, Jeffrey Bub, and Hans Halvorson [15]. In that paper, Clifton et al. write:

> The foundational significance of our derivation, as we see it, is that quantum mechanics should be interpreted as a *principle theory*, where the principles at issue are information-theoretic.

Since Brown and Timpson already believe that special relativity needs a constructive counterpart, they certainly do not advocate a principle approach to quantum mechanics and they base their argument on the constructive completion of thermodynamics.

As Einstein pointed out above, thermodynamics is a principle theory based on the principle that perpetual motion is impossible. For example, a perpetual motion machine of the second kind violates the second law of thermodynamics, i.e., entropy always increases. Let's look at that.

The change in entropy ΔS_S for some system is equal to the amount of heat it gains, ΔQ_S, divided by its temperature T_S. The change in entropy for the environment of that system, ΔS_E, in this process would then be equal to the amount of heat it gains, ΔQ_E, divided by its temperature T_E. Since the amount of heat gained (lost) by the system equals the amount of heat lost (gained) by the environment, we have $\Delta Q_S = -\Delta Q_E$, so

$$\Delta S_{\text{Total}} = \Delta S_S + \Delta S_E = \frac{\Delta Q_S}{T_S} - \frac{\Delta Q_S}{T_E} = \left(\frac{1}{T_S} - \frac{1}{T_E}\right)\Delta Q_S.$$

The second law says ΔS_{Total} must be greater than zero (it must increase), so if ΔQ_S is positive (the system gains heat from the environment), then it must be the case that $T_S < T_E$. Conversely, if ΔQ_S is negative (the system loses heat to the environment), then it must be the case that $T_E < T_S$.

That is, heat must flow spontaneously from high temperature to low temperature. If someone designs a perpetual motion machine based on heat flowing spontaneously from low temperature to high temperature, then we can say it is ruled out by the second law of thermodynamics. Likewise, a perpetual motion machine of the first kind violates the first law of thermodynamics (conservation of energy), and a perpetual motion machine of the third kind violates the third law of thermodynamics (a heat engine cannot achieve 100% efficiency).

Conversely, one could say that thermodynamics is based on the empirically discovered fact that perpetual motion is impossible, so that:

no perpetual motion of Xth kind → Xth law of thermodynamics.

That is the way Einstein viewed thermodynamics. Brown and Timpson point out that thermodynamics provides "phenomenological laws which stipulate nothing about the deep structure of the working substance" like we get from the constructive kinetic theory of matter. There, we understand that the high-temperature environment contains particles of higher kinetic energy on average than the particles of the lower-temperature system. And, as the faster particles strike the slower particles, the slower particles speed up a bit while the faster particles slow down a bit. That is just in accord with conservation of momentum in a collision process and entails that heat flows from high temperature to lower temperature. So, just like with special relativity, the constructive account reverses the explanatory hierarchy of the principle account:

| constructive kinetic | → | Xth law of | → | no perpetual motion |
| theory of gases | | thermodynamics | | machines of the Xth kind |

Clearly, Brown and Timpson argue, the constructive explanation is preferred over the principle explanation here, and we would certainly agree.

Likewise, as we will see in Chapter 6, the information-theoretic reconstructions of quantum mechanics are based on the empirically discovered fact of complementarity/superposition/noncommutativity. So, Brown and Timpson argue [8]:

> It is a remarkable thing that what might be called the kinematic structure of quantum theory, the nature of its observables and state space structure, can it seems be given a principle-theory, or 'thermodynamic' underpinning. As Bell stressed, the beauty of thermodynamics is in its economy of reason, but the insight it provides is limited in relation to the messier story told in statistical mechanics.

If that was the end of the story, we would agree with Brown and Timpson that quantum mechanics as a principle theory based on the information-theoretic counterpart to complementarity/superposition/noncommutativity cries out for completion, just like thermodynamics based on the impossibility of perpetual motion. And, the compelling constructive completion of thermodynamics via statistical mechanics does suggest we look for a constructive completion of quantum mechanics. However, the story does not end here for quantum mechanics or thermodynamics.

That is, by showing how the relativity principle justifies the empirically discovered fact at the foundation of information-theoretic reconstructions of quantum mechanics (NPRF + h), we see that quantum mechanics provides a principle *explanation* as compelling as special relativity (NPRF + c). And, both theories are without a (consensus) constructive completion after many decades of effort. The addition of a compelling fundamental principle (NPRF) that justifies the empirically discovered facts (the Planck and light postulates) is what sells quantum mechanics and special relativity as principle explanations. [Recall our distinction between a principle theory and a principle explanation in Chapter 1.]

Likewise, thermodynamics as a principle theory based on the empirically discovered fact of "no perpetual motion" needs to be completed via a compelling fundamental principle to qualify as a principle explanation. Ironically, it finds that justification via its constructive completion. That is, since statistical mechanics uses Newtonian mechanics and Newtonian mechanics is merely an approximation to quantum mechanics and special relativity, we see that the ultimate (most foundational) completion of thermodynamics is not constructive, but principle. And to exacerbate the irony, that ultimate compelling fundamental principle is ... NPRF (Figure 5.1).

While Einstein did have a clear preference for constructive explanation, he did also have a clear appreciation for any explanation that offered extensive unification. Einstein writes [17, p. 32]:

> A theory is the more impressive the greater the simplicity of its premises is, the more different kinds of things it relates, and the more extended its area of applicability.

Note that in the same quote where he complains about special relativity's lack of explanatory power in his letter to Sommerfeld, he laments the "vain efforts to interpret" Planck's constant h. We are not historians, but we suspect that Einstein hoped a complete version of quantum mechanics would produce an acceptable constructive counterpart to special relativity, thereby extending the principle unification of mechanics and electrodynamics (per special relativity) to a constructive unification of mechanics, electrodynamics, *and* quantum mechanics.

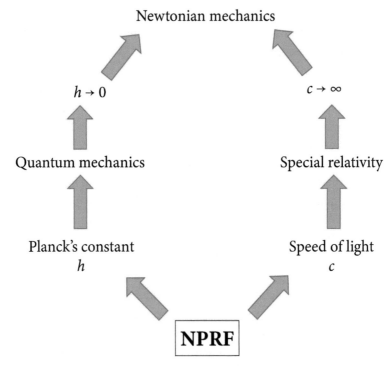

Fig. 5.1 The variables in Newtonian mechanics commute, which means $h \to 0$ in the commutator for the corresponding variables in quantum mechanics. Newtonian equations hold on average according to quantum mechanics and follow from the corresponding equations in special relativity with $c \to \infty$.

It only makes sense because quantum mechanics is a theory of matter fundamental to classical physics, and meter sticks and clocks are made of matter, so it would seem that length contraction and time dilation should be explicable most fundamentally by quantum mechanics. But in order for that to happen, the complete version of quantum mechanics would have to be constructive, and per Bell's theorem, as we showed in Chapters 2 and 3, we now know that a constructive account of quantum mechanics is problematic. Nonetheless, Einstein's constructive bias is still echoed today.

For example, in his book *Quantum Non-Locality and Relativity* [25] Maudlin makes it clear that he accepts Reichenbach's Principle as necessary for the explanation of EPR–Bell correlations, since he only discusses constructive accounts of EPR–Bell correlations. The constructive bias as regards the behavior of matter–energy is understandably very strong (see Section 3.4), but nonetheless no one has done a better job pointing out the fatal flaws of all extant constructive accounts than Maudlin in that book. Regarding his own attitude about these constructive accounts, Maudlin simply says:

> I have not advocated a single choice among these options. It is very diffi-cult to imagine a clear set of uncontroversial standards of plausibility that would yield a definitive decision here. Perhaps other, more unusual and counterintuitive possibilities may be articulated.

Indeed, the constructive bias is so strong that some philosophers, such as Brown [7, 9], have argued for our hypothetical defense of Einstein above. As Brown puts it [7, pp. vii–viii]:

> In a nutshell, the idea is to deny that the distinction Einstein made in his 1905 paper between the kinematical and dynamical parts of the discussion is a fundamental one, and to assert that relativistic phenomena like length contraction and time dilation are in the last analysis the result of structural properties of the quantum theory of matter.

But, when one truly appreciates the principle account of special relativity it becomes clear that length contraction is not about the mechanical or construc-tive shrinking of meter sticks. Length contraction is a relative, kinematic effect according to the relativity of simultaneity as necessitated by NPRF + c (as we explained in Chapter 4). As it turns out, Maudlin's constructive bias does not run so deep as Brown's and he happily accepts and promotes the kinematic account of special relativity [25, pp. 48–53], as with most physicists today. So, given that:

- special relativity is widely accepted as a principle theory with no con-structive counterpart;
- a constructive account of quantum mechanics necessarily violates local-ity, statistical independence, intersubjective agreement, or unique exper-imental outcomes (in order to solve the mystery of entanglement);
- we now have a principle account of quantum mechanics based on the relativity principle à la special relativity (Chapters 6 and 7);

is it not reasonable to base quantum mechanics on Einstein's principle rather than Reichenbach's? At the very least, such a principle account would serve as a significant step towards understanding quantum mechanics. That is what we will argue in the remainder of this chapter.

Pointing to an historical trend from the 19th century, Marco Giovanelli notes that [20]:

... despite its apparent radical novelty, the relativity principle, like the energy principle, is ultimately an instance of "that general direction of physical thought, which has been called the 'physics of principles' in contrast to the physics of pictures and mechanical models." [Quote from E. Cassirer, *Zur Einstein'schen Relativitätstheorie*]

and [20]:

The initial contradiction between mechanics and electrodynamics that is revealed by the negative result of ether drift experiments was overcome not "by using the electrodynamic processes as a key to the mechanical" but by establishing "a far more perfect and deeper unity between the two than previously existed" ... The unification of the two separate fields of theoretical physics—electrodynamics and mechanics—is not obtained through a process of horizontal integration, a reduction of the one to the other, but through a vertical integration, a subsumption of both theories under a higher principle. [Quote from E. Cassirer, *Zur Einstein'schen Relativitätstheorie*]

In short, the reason "nobody understands quantum mechanics" is precisely because the foundations community is wedded to "the physics of pictures and mechanical models" used for "deriving the principles from what are believed to be the laws of nature" (Section 3.4), rather than testing "the acceptability of the laws of nature through certain general principles" [20]. According to our proposal for understanding quantum mechanics, to co-opt Giovanelli [20]:

The unification of the two separate fields of theoretical physics—special relativity and quantum mechanics—is not obtained through a process of horizontal integration, a reduction of special relativity to quantum mechanics, but through a vertical integration, a subsumption of both theories under the relativity principle.

And this is made possible by completing the reconstruction program as a principle explanation (Chapter 6).

As we stated in Chapter 1, you do not hear Nobel Laureates saying "nobody understands special relativity," even though it is a principle theory without a consensus constructive counterpart. That is because the mysterious empirically discovered fact postulated by special relativity (the light postulate) is justified by NPRF, a principle that physicists find compelling. So, by showing how the mysterious empirically discovered fact postulated by information-theoretic reconstructions of quantum mechanics (the Planck postulate) is justified by the very same relativity principle should certainly constitute a significant step towards "understanding quantum mechanics."

5.2 Newtonian Mechanics from Quantum Mechanics and Special Relativity

Conventional thinking is that quantum mechanics and special relativity are more fundamental than Newtonian mechanics, and we agree. That is, the equations of Newtonian mechanics hold on average according to quantum mechanics, e.g., as with conservation of spin in Chapter 0, and they follow from letting $c \to \infty$ in special relativistic mechanics. For example, relativistic kinetic energy is given by

$$\text{KE} = mc^2 \left(\frac{1}{\sqrt{1 - v^2/c^2}} - 1 \right).$$

If $v \ll c$, which is like letting $c \to \infty$, we have

$$\text{KE} = mc^2 \left(\frac{1}{\sqrt{1 - v^2/c^2}} - 1 \right) \approx mc^2 \left(\left(1 + \frac{v^2}{2c^2} \right) - 1 \right) = \frac{1}{2}mv^2,$$

which you will recognize as the Newtonian kinetic energy equation.

As we stated in Chapters 0 and 1, and will show in Chapters 6 and 7, quantum mechanics follows from the relativity principle and Planck postulate, and as we explained in Chapters 1 and 3, it has no consensus constructive account. Likewise, as we explained in Chapter 4, special relativity follows from the relativity principle and light postulate, and it has no consensus constructive account. Therefore, Newtonian mechanics is based most fundamentally on NPRF + h and NPRF + c with no *fundamental* constructive account (Figure 5.1).

So, is it reasonable to impose our intuitive dynamical/constructive bias as developed in Newtonian mechanics on the more fundamental quantum mechanics? No, it is completely backward, we need to impose the principle approach to quantum mechanics on Newtonian mechanics because we have no (acceptable) constructive account of quantum mechanics, and Bell's theorem assures us none is forthcoming. That means we need to look at how the fundamental principle approach to quantum mechanics translates to New-tonian mechanics, only then can we can identify situations where the less fundamental dynamical/constructive approach is justified. To do that, we need to return again to basic physics.

In both Newtonian mechanics and special relativity (classical mechanics), one starts with kinematics, i.e., the physics that can be done using only posi-tion, velocity, and acceleration (Table 5.1). In those theories, as we explained in Chapters 1 and 4, the kinematics is dictated by the relativity principle, i.e., Galileo's version for Newtonian mechanics and Einstein's version for special relativity. And, the dynamics is constrained by the kinematics, i.e., Newto-nian dynamics is invariant under Galilean transformations and relativistic dynamics is invariant under Lorentz transformations.

We have a similar situation for quantum mechanics, but its kinematics has to do with probability theory and is represented by the probability structure of Hilbert space [12], as we described in the preface and used in Chapters 2 and 3. In quantum mechanics, as we will see in Chapters 6 and 7, the kinematics is also dictated by the relativity principle, i.e., the empirically dis-covered fact from which one builds Hilbert space is Information Invariance & Continuity and it is justified by NPRF. The dynamics of the quantum-mechanical probability theory is given by Schrödinger's equation, which gov-erns the time evolution of the wavefunction (quantum state) in Hilbert space (Table 5.1).

What we find from the kinematic structure of quantum mechanics is that the equations in Newtonian mechanics using commuting variables hold on average over the noncommuting variables of quantum mechanics. That is how the kinematics of quantum mechanics relates to the dynamics of Newtonian mechanics. Let's clarify this using the Mermin device.

When we are talking about conservation of spin angular momentum per Fact 1 for case (a), i.e., Alice and Bob obtain the same outcomes RR or GG whenever they make the same measurement, the Bell state of quantum

Table 5.1 Comparison of classical and quantum-mechanical kinematics and dynamics. Classical mechanics consists of special relativity and Newtonian mechanics.

Theory	Kinematics	Dynamics
Classical mechanics	Position, velocity, acceleration	Momentum, force, energy, etc.
Quantum mechanics	Hilbert space	Schrödinger's equation

mechanics gives the Newtonian conservation principle exactly, no averaging needed. That is because Alice and Bob are making the same measurement in case (a), so there is no concern about noncommutativity, i.e., Bob's (Alice's) spin measurement outcome allows him (her) to simultaneously know the value of Alice's (Bob's) spin angular momentum in the same direction (Section 3.2). In that case, we can easily construct a dynamical counterpart to the Bell state conservation without violating locality or statistical independence by using instruction sets, i.e., we can invoke the hidden variables missing from quantum mechanics. All is well for constructive explanation because we can again subordinate the conservation principle (Bell state conservation) to the dynamical law (represented by the hidden variables).

But, when we are talking about conservation of spin angular momentum per Fact 2 for case (b), i.e., Alice and Bob's measurement outcomes agree in $\frac{1}{4}$ of the trials when their measurement settings differ, quantum mechanics only gives the Newtonian conservation principle *on average*. Indeed, the case (b) conservation is 'average-only,' meaning it *never* holds exactly as given by Newtonian mechanics in any given trial of the experiment.[1] That is because the Planck postulate per NPRF (Chapters 0 and 2) dictates that Bob's (Alice's) spin measurement outcome does not allow him (her) to simultaneously know the value of Alice's (Bob's) spin angular momentum in a different direction (Section 3.2), i.e., spin angular momentum in different directions does not commute. So, let's put all of this together.

The quantum state (Bell state) represents the conservation of spin angular momentum for the Mermin device in accord with NPRF. It is fundamental, i.e., there is no dynamical/constructive equation in the formalism of quantum mechanics from which one derives that Bell state for the conservation of spin angular momentum as with Eq. (3.5) in Newtonian mechanics. Whether or not the conservation principle is more fundamental than the dynamical law for quantum mechanics in this case is a nonstarter because there is no dynamical law. That is why Einstein complained that quantum mechanics is incomplete. Further, you cannot even construct a dynamical law that accounts for Fact 2 for case (b) without violating locality, statistical independence, intersubjective agreement, or unique experimental outcomes, which most deem necessary for proper dynamical/constructive explanation. Again, that is how the mystery of

---

[1] There is no way to save trial-by-trial conservation for case (b) à la the work of Yakir Aharonov, Sandu Popescu, and Daniel Rohrlich [3, 4] because it is neither Bob nor Alice's results that are violating strict conservation, it is purely relative, so there is nothing 'missing' or 'excessive' that needs to be found or accounted for.

entanglement is based on dynamical/constructive bias. Now let's return to our question, i.e., is that bias justified?

We do not see how. Again, everyone agrees that quantum mechanics is fundamental to Newtonian mechanics, and the conservation principle in question for quantum mechanics (Bell state) is trivially more fundamental than a nonexisting dynamical counterpart. That is, unlike Newtonian mechanics, Bell state conservation has no dynamical counterpart in the formalism of quantum mechanics and a consensus-acceptable dynamical law is provably impossible. So, if one is to infer which is more fundamental in the corresponding Newtonian mechanics, conservation principle or dynamical law, given the situation in the more fundamental quantum mechanics, the favorite would have to be the conservation principle.

And, since the Bell state conservation principle is justified by NPRF, it looks to us like Einstein's relativity principle is indeed more fundamental than Reichenbach's Principle for solving the mystery of entanglement. That is because Einstein's relativity principle holds for both Facts 1 and 2 while Reichenbach's Principle holds for Fact 1, but it can only hold for Fact 2 by violating locality, statistical independence, intersubjective agreement, or unique experimental outcomes.

5.3 Can We Save Reichenbach in Spirit?

Clearly, Bell's theorem and the experimental violation of Bell's inequality tell us that Reichenbach's Principle fails for the correlations of entanglement, at least if one wants causality along the lines of Newtonian force and torque (Section 3.4) in accord with locality, statistical independence, intersubjective agreement, and unique experimental outcomes. But, restricting the notion of causality to Newtonian force and torque is a rather simplistic view of causality. Plus, again, Newtonian mechanics is not a fundamental theory of physics. As Adlam stated [1]:

> there have been several major conceptual revolutions in physics since the time of Newton, and thus we should not necessarily expect that accounts of lawhood based on a Newtonian time-evolution picture will be well-suited to the realities of modern physics.

Perhaps there are more sophisticated ways to understand causality in accord with the more fundamental theory of quantum mechanics, so as to salvage

Reichenbach's Principle in some sense. For example, as we pointed out in Chapter 3, Wharton and Liu [39] used the path integral formalism[2] for quantum mechanics that "is much more evidently time-symmetric than conventional [quantum mechanics], bringing it closer to a causally-neutral account." Let's look more closely at causation per their all-at-once retrocausality.

For the Mermin device, all-at-once retrocausality is often described as making a timelike "zigzag" pattern in spacetime between the two spacelike separated measurement events [30, 31, 40], as shown in Figure 5.2. As we stated in Chapter 3, the main complaint about timelike zigzag causation is that events are not being brought into existence in accord with our everyday intuition because part of the zigzag causal influence is coming from future events that have never existed. But, again, the belief that those future events have never existed is in accord with the Newtonian objective spacetime model of reality, which is challenged on the coexistence of bodily objects and events by the

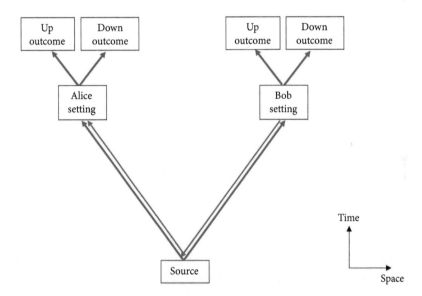

Fig. 5.2 The blue arrows depict two particles leaving the source towards Alice and Bob's Stern–Gerlach (SG) measurement devices. Each trial of the experiment produces one of the two outcomes shown above each of their settings. The red arrows depict a timelike causal "zigzag" pattern from Bob's SG measurement setting to the source then to Alice's SG measurement outcome.

---

[2] We will explain the path integral formalism a bit more in Chapter 9, but you do not need to know exactly what it is to follow the argument here.

objective spacetime model of reality for special relativity. So, let's consider a notion of causality that is not necessarily Newtonian in that sense.

In Figure 5.2, for example, Bob's Stern–Gerlach setting creates a causal influence backward in time to the source then forward in time to causally influence the particle–detector interaction at Alice's SG setting. If Alice and Bob's setting events were timelike related, one could simply invoke a forward-time-directed causal influence from the earlier event directly to the later event. Since Alice and Bob's setting events are spacelike related, a zigzag causal influence is needed and the opposing temporal directions of that influence mean it is all-at-once or atemporal. No events are *objectively* bringing other events into existence. This is consistent with the fact that we could have drawn the zigzag from Alice's SG setting to Bob's and none of the empirical facts would have 'changed' in our all-at-once zigzag diagram.

This symmetry distinguishes forward timelike causation from retrocausation and is directly related to quantum phenomena. To see why that is true, consider a different experiment altogether whereby a single particle is input to Bob's SG magnets in one of two states corresponding to up or down for his particular SG magnet setting. Now we are sending a particle to Bob's SG magnets with spin up (or down) relative to Bob's SG magnet setting, so of course the particle merely proceeds through his SG magnets unchanged to Alice's SG magnets (Figure 5.3). We can think of this single-particle experiment as resulting from temporally flipping Bob's half of the experimental trial shown in Figure 5.2 for two particles to produce Figure 5.3 for one particle.

For example, suppose we send a spin up particle to Bob's SG magnets so that it proceeds as a spin up particle with respect to Bob's SG magnets to Alice's SG magnets. There, the particle emerges either up (+1) or down (−1) with respect to Alice's SG magnets exactly as given by the quantum-mechanical probability, i.e., $P(+ \mid \theta) = \cos^2(\theta/2)$ and $P(- \mid \theta) = \sin^2(\theta/2)$, where θ is the angle between Bob's SG magnets and Alice's (as we showed in Section 3.2). The situation is analogous when we start with a spin down particle, we just have to exchange sine and cosine in the probability functions. In this purely forward-timelike sequence of events, we have no trouble saying it is Bob who establishes the cause, i.e., the initial quantum state for Alice to measure. Certainly, there is nothing mysterious here (other than the mystery of spin itself, as we explained in Chapter 2).

Next, because the zigzag pattern can go the other way, consider reversing the zigzag pattern and then flipping Alice's half of the experimental trial shown in Figure 5.2 for two particles to produce Figure 5.4 for one particle. Everything we said about Bob can now be said about Alice and vice versa. That includes

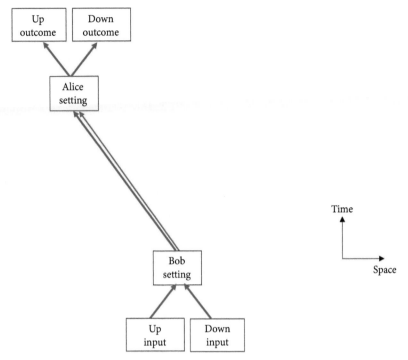

Fig. 5.3 We temporally flip Bob's side of a trial for the Mermin device from that in Figure 5.2. Now there is only one particle moving forward in time from either of two inputs to Bob's SG magnets then to Alice's SG magnets where we have one of two possible outcomes.

the fact that in the temporally flipped case it is clearly Alice who establishes the cause, i.e., the initial quantum state for Bob to measure. So, how is this symmetry due to the quantum nature of spin?

If the classical model of spin were true (Figure 2.2), then Alice or Bob would be measuring a fraction of h when their SG magnets were not aligned. In that case, Alice or Bob's fractional outcome definitely establishes an asymmetry that does not allow for this symmetrical temporal flipping. That is, if Bob's (Alice's) outcome is a fraction, then he (she) has to follow Alice's (Bob's) state preparation in a temporally flipped scenario because you cannot restore Bob's (Alice's) missing spin angular momentum with a subsequent measurement. Notice that this asymmetry results from violating NPRF.

Because NPRF holds, Planck's postulate holds and we have the symmetrical quantum outcomes, so Alice and Bob's choices for their SG magnet settings carry equal causal weight. That is, instead of Bob's choice of his SG magnet

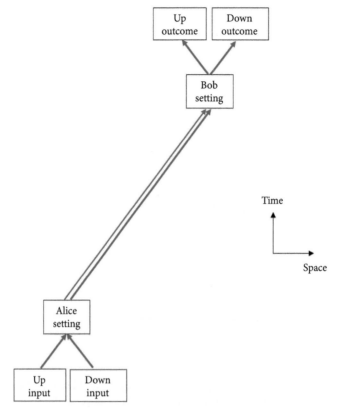

Fig. 5.4 We temporally flip Alice's side of a trial for the Mermin device from that in Figure 5.2 (after flipping the zigzag red arrows). Now there is only one particle moving forward in time from either of two inputs to Alice's SG magnets then to Bob's SG magnets where we have one of two possible outcomes.

setting causing the quantum state for Alice to measure or vice versa, either SG magnet setting can be said to cause the quantum state for the other. Recall Figures 0.4 and 0.6 and what we said about them in Chapter 0 regarding the symmetrical 'average-only' conservation. There, Alice partitioned the data according to her +1 and −1 outcomes and said Bob needed to average his results to satisfy conservation of spin, while Bob partitioned the data according to his +1 and −1 outcomes and said Alice needed to average *her* results to satisfy conservation of spin.

Zigzag causation is symmetric or co-causation and since it arises from a symmetry due to the quantum, some people call it time-symmetric causation. In addition to time-symmetric causation, there are also those who invoke indefinite causal ordering [5, 6, 32]. Natalie Wolchover writes [41]:

Over the last decade, quantum physicists have been exploring the implications of a strange realization: In principle, both versions of the story can happen at once. That is, events can occur in an indefinite causal order, where both "A causes B" and "B causes A" are simultaneously true.

Some wavefunction realists are inclined to take such claims literally, whatever that may mean. The point is that many theorists will go to great lengths to stretch the notion of causation in order to keep causal explanation rather than fully adopt an all-at-once approach. As pointed out by Maudlin, whether we are talking about time-symmetric causation, retrocausation, indefinite causal ordering, etc., this violates the essence of causal explanation as standardly conceived [26]:

> This is a distinction that all physicists—including Newton and Einstein—as well as all everyday folk take for granted. The before/after distinction grounds the cause/effect distinction: everyone would accept that an earlier configuration of the planets, together with the laws of gravity, cause and explain their later configuration, but no one would say the later configuration causes or explains the earlier. It may *indicate* or *allow one to infer* the earlier, but not *explain* it.

This is all in keeping with our classical physics bias developed in accord with our everyday dynamical experience. Again, in the classical model of conservation, we would have a clear asymmetry between Alice and Bob's outcomes allowing us to pick a preferred reference frame and definitive causal order. But as we showed, our Bell state conservation is in accord with NPRF and it only produces classical physics on average, i.e., quantum mechanics is fundamental to Newtonian mechanics. So if we accept 'average-only' conservation as the more fundamental form of conservation, should we also be willing to accept strange accounts of quantum causality as the more fundamental form of causality?

One can certainly provide accounts of causation that are consistent with the weirdness of the quantum, such as interventionist or manipulability accounts of causation [27]. The central idea is that X is a cause of Y if and only if manipulating X is an effective means of indirectly manipulating Y. According to retrocausal accounts of quantum mechanics espousing an interventionist account of causation, manipulating the setting of a measurement apparatus now can be an effective means of manipulating aspects of the past. The formal machinery of causal modeling has the interventionist account of causality as its foundation. For the Mermin device, that means the Bell state correlations depend on Bob's SG settings, so Bob has intervened in these correlations. Likewise, the Bell state correlations depend on Alice's SG settings, so Alice has intervened in these correlations.

Price and Wharton, two key defenders of retrocausal accounts of the quantum, embrace a subset of interventionism known as the agent or perspectivalism account of causation [19, 28, 29]. On this view, causal relations are relations that can be used for control or manipulation, from the perspective of the agent in question of course. This is an understandably appealing notion of causation for those such as Price and Wharton who espouse a block universe picture, wherein causation talk cannot possibly be about changing or bringing about events (past, present, or future) in any robust sense of those terms. So, such accounts of causation do not provide physical explanations for EPR–Bell correlations, they just attempt to save the appearances of causal explanations.

5.4 Co-causation or No Causation?

For example, if we think of Figure 5.2 as Bob's subjective spacetime model of reality according to perspectivalism and the flipped zigzag version as Alice's subjective spacetime model of reality, then what happens to the zigzag in the objective spacetime model of reality? The combined effect of both zigzags amounts to what Wharton and Liu called a "causally-neutral account," so the zigzags are functionally superfluous in the objective spacetime model of reality. By simply omitting the zigzags in the objective spacetime model of reality, we could avoid causal talk altogether and conclude that NPRF + h provides an all-at-once AGC explanation of Bell state entanglement that does not violate locality, statistical independence, intersubjective agreement, or unique experimental outcomes. It is only when we attempt to further underwrite NPRF + h via zigzag paths in spacetime that we introduce constructive explanation and violate locality and/or statistical independence. It is the desire to keep constructive explanation, causal explanation, etc. at all costs that gets us into trouble when explaining EPR–Bell correlations.

To see that, consider Price and Wharton's recent approach to underwriting all-at-once retrocausality using "connections across a constrained collider" [31]. A collider is a point where causal influences converge (Figure 5.5). In their simple example, patients in Ward C are infected with a rare virus A and/or a rare virus B, so being infected with either of the two viruses is a causal influence for being a patient in Ward C. This introduces classical entanglement

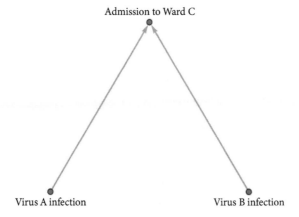

Fig. 5.5 A simple collider [31].

between the patients in Ward C, i.e., if someone is in Ward C and they do not have virus A, you know immediately that they have virus B, and vice versa.

We can think of the Mermin device as a collider by temporally flipping the causal diagram to create a past collider (Figure 5.6). Now our (retro)causal influences are Alice and Bob's choices of measurement settings converging to the source of the entangled Bell state (represented by Charlie's selection of outcomes for Alice and Bob's future measurements). Again, using the analysis above, Alice can say that her measurement choice influences Bob's measurement outcome in zigzag fashion through the source per Charlie's selection of outcomes (Figure 5.7) while the zigzag is reversed from Bob's perspective.

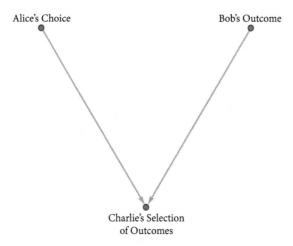

Fig. 5.6 A past collider [31].

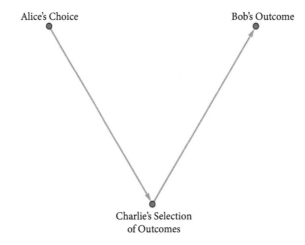

Fig. 5.7 Past collider with a zigzag [31].

So, we have violated statistical independence by trying to explain the all-at-once retrocausality of the objective spacetime model of reality as the result of two timelike causal zigzag paths from the two subjective spacetime models of reality according to perspectivalism. Notice the analogy with length contraction and time dilation in Chapter 4.

There, from Alice's perspective (her reference frame), Bob's meter sticks are short and his clocks run slow, while from Bob's perspective (his reference frame) Alice's meter sticks are short and her clocks run slow. Here, Alice can claim it is her choice of measurement setting that is causally influencing Bob's measurement outcome (Figure 5.7), while Bob can say the same thing about his choice of measurement setting causally influencing Alice's measurement outcome. Is it better to think of this as quantum co-causation? Or, should we rather think of it as no causation? Let's look again at special relativity to help us answer that question.

If Alice's perspective (reference frame) is preferred, then there must be some causal mechanism literally shortening Bob's meter sticks and slowing down his clocks. As a result, Bob measures the same value c for the speed of a light beam that Alice does, even though they are moving relative to each other. Conversely, if Bob's perspective (reference frame) is preferred, the causal mechanism is actually shortening Alice's meter sticks and slowing down her clocks, again resulting in her measuring the same value c for the speed of a light beam that Bob does. What is the consensus attitude about this in physics?

There is no causal mechanism shortening meter sticks and slowing down clocks in any reference frame. Alice's measurements indicate that Bob's meter sticks are short and his clocks run slow, while Bob's measurements indicate the same thing about Alice's meter sticks and clocks, precisely because neither perspective is preferred (NPRF). The relativity principle means Alice and Bob must both measure the same value c for the speed of a light beam, and that leads to the relativity of simultaneity which explains Alice's and Bob's measurement outcomes without having to invoke causal mechanisms at all. It is all about principle explanation rather than constructive explanation.

Extrapolating the principle approach from special relativity to the quantum situation here suggests we again drop the notion of different causal influences invoked from different perspectives, i.e., different causal zigzag patterns invoked from their different reference frames. If there is some causal mechanism responsible for their disagreement, it is responsible for Bob measuring the same value h for Planck's constant that Alice does, even though their measurement devices are oriented in different directions.

So, following the analogous situation in special relativity, we might simply adopt the relativity principle to justify the empirical fact that Alice and Bob both measure the same value h for Planck's constant. As we show in Chapters 6 and 7, that leads to 'average-only' conservation which explains Alice's and Bob's measurement outcomes without having to invoke causal mechanisms at all. Consequently, this NPRF + h principle explanation does not violate locality, statistical independence, intersubjective agreement, or unique experimental outcomes. Perhaps we should abandon constructive causal accounts and opt for all-at-once adynamical global constraints?

5.5 Summary

If we want to solve the mystery of entanglement in accord with Reichenbach's Principle, so as to resolve the EPR and EPR–Bell paradoxes and answer Mermin's challenge constructively, then we must violate locality, statistical independence, intersubjective agreement, and/or unique experimental outcomes. This leads to a problem articulated by Van Camp [38]: "Constructive interpretations are attempted, but they are not unequivocally constructive in any traditional sense." He concludes [38]:

> The interpretive work that must be done is less in coming up with a constructive theory and thereby explaining puzzling quantum phenomena, but more in explaining why the interpretation counts as explanatory at all given that it must sacrifice some key aspect of the traditional understanding of causal-mechanical explanation.

If statistical mechanics is the paradigm example of constructive explanation, then in light of Bell's theorem and the experimental violation of Bell's inequality, it is hard to imagine any constructive account of quantum mechanics gaining consensus support. And, if special relativity (following from NPRF + *c*) is the paradigm example of principle explanation without a consensus constructive counterpart, it is easy to imagine quantum mechanics (following from NPRF + *h*) also remaining without a consensus constructive counterpart.

As we pointed out in Chapter 3, there is already a definite trend towards atemporal / all-at-once / AGC explanation for Bell state correlations. Adlam and Rovelli write [2]:

> Moreover, ... we do not need to think of the set of events as being generated in some particular temporal order. In fact, we can say something even stronger: if we want to maintain relativistic covariance then we cannot think of the set of events as being generated in some particular temporal order. This point has been noted in the context of other ontologies consisting of point-like events—for example, Esfeld and Gisin note that the Bell flash ontology is relativistically covariant only if "one limits oneself to considering whole possible histories or distributions of flashes in spacetime, and one renounces an account of the temporal development of the actual distribution of the flashes in space-time." ... Thus it seems that [relational quantum mechanics] is most compatible with a metaphysical picture in which the laws of nature apply atemporally to the whole of history, fixing the entire distribution of quantum events all at once.

They note that this kind of explanation does not "involve hidden influences or preferred reference frames, and thus there is no particular reason to try to avoid this sort of nonlocality."[3] As we showed in Chapter 3 and above, Hance, Hossenfelder, Palmer, Price, Wharton, and Liu have also adopted all-at-once explanation for superdeterminism, supermeasured theory, and retrocausality. Concerning the laws of physics more generally, Eddy Chen and Sheldon Goldstein write [14]:

> It is sometimes assumed that the governing view of laws requires a fundamental direction of time: to govern, laws must be *dynamical* laws that *produce*

---

[3] Here, "nonlocality" is referring to relations between events distributed over large regions of spacetime as opposed to small (local) regions of spacetime. To avoid confusion with nonlocality per superluminal causal connections, the better term for this type of "nonlocality" is probably "nonseparability" [34].

later states of the world from earlier ones, in accord with the direction of time that makes a fundamental distinction between past and future. ... On our view, fundamental laws govern by constraining the physical possibilities of the entire spacetime and its contents. They need not exclusively be dynamical laws, and their governance does not presuppose a fundamental direction of time. For example, they can take the form of global constraints or boundary-condition constraints for spacetime as a whole;

Since this is precisely how one should view the principle constraint NPRF + *h*, it is perhaps time to reject Reichenbach's Principle altogether in favor of Einstein's for solving the mystery of entanglement specifically and understanding quantum mechanics more generally. At the very least, NPRF + *h* should not be dismissed as an unreasonable solution to the mystery of entanglement. We hope Chapters 0–5 have convinced you of that fact.

If so, you are in the proper state of mind to see how NPRF + *h* is obtained from the so-called information-theoretic reconstruction of quantum mechanics. As we will show, NPRF + *h* has a well-developed history in the information-theoretic reconstructions of quantum mechanics. For example, Bub and Pitowski have already pointed out the analogy between Minkowski spacetime and Hilbert space [11, 12, 13] in an attempt to explain Bell state correlations. As Bub sums it up [10]:

Hilbert space as a projective geometry (i.e., the subspace structure of Hilbert space) represents the structure of the space of possibilities and determines the kinematic part of quantum mechanics. ... The possibility space is a non-Boolean space in which there are built-in, structural probabilistic constraints on correlations between events (associated with the angles between the rays representing extremal events) – just as in special relativity the geometry of Minkowski space-time represents spatio-temporal constraints on events. These are kinematic, i.e., pre-dynamic, objective probabilistic or information-theoretic constraints on events to which a quantum dynamics of matter and fields conforms, through its symmetries, just as the structure of Minkowski space-time imposes spatio-temporal kinematic constraints on events to which a relativistic dynamics conforms.

In the next two chapters we will unpack that information-theoretic language and move this idea beyond a mere analogy by showing how both Hilbert space and Minkowski spacetime are based on the relativity principle. Specifically, quantum mechanics is based on NPRF + h, as can be seen using Information Invariance & Continuity from the information-theoretic reconstructions of quantum mechanics.

References

[1] E. Adlam, *Laws of Nature as Constraints*, 2021. Preprint. https://arxiv.org/abs/2109.13836.

[2] E. Adlam and C. Rovelli, *Information is Physical: Cross-Perspective Links in Relational Quantum Mechanics*, 2022. Preprint. https://arxiv.org/abs/2203.13342.

[3] Y. Aharonov, S. Popescu, and D. Rohrlich, *On Conservation Laws in Quantum Mechanics*, Proceedings of the National Academy of Sciences, 118 (2021), p. e1921529118.

[4] Y. Aharonov, S. Popescu, and D. Rohrlich, *Conservation laws and the foundations of quantum mechanics*, Proceedings of the National Academy of Sciences, 120 (2023), p. e2220810120.

[5] M. Araújo, F. Costa, and C. Brukner, *Computational Advantage from Quantum-Controlled Ordering of Gates*, Physical Review Letters, 113 (2014), p. 250402.

[6] V. Baumann, M. Krumm, P. A. Guérin, and C. Brukner, *Noncausal Page–Wootters circuits*, Physical Review Research, 4 (2022), p. 013180.

[7] H. Brown, *Physical Relativity: Spacetime Structure from a Dynamical Perspective*, Oxford University Press, Oxford, 2005.

[8] H. Brown and C. Timpson, *Why special relativity should not be a template for a fundamental reformulation of quantum mechanics*, in Physical Theory and Its Interpretation: Essays in Honor of Jeffrey Bub, W. Demopoulos and I. Pitowsky, eds., Springer, New York, 2006.

[9] H. R. Brown and J. Read, *The dynamical approach to spacetime theories*, in The Routledge Companion to Philosophy of Physics, E. Knox and A. Wilson, eds., Routledge, New York, 2021.

[10] J. Bub, *Quantum Correlations and the Measurement Problem*, 2012. Preprint. https://arxiv.org/abs/1210.6371.

[11] J. Bub, *Bananaworld: Quantum Mechanics for Primates*, Oxford University Press, Oxford, 2016.

[12] J. Bub, *"Two Dogmas" Redux*, in Quantum, Probability, Logic: The Work and Influence of Itamar Pitowski, M. Hemmo and O. Shenker, eds., Springer Nature, London, 2020, pp. 199–215.

[13] J. Bub and I. Pitowski, *Two dogmas about quantum mechanics*, in Many Worlds? Everett, Quantum Theory, and Reality, S. Saunders, J. Barrett, A. Kent, and D. Wallace, eds., Oxford University Press, Oxford, 2010, pp. 431–456.

[14] E. Chen and S. Goldstein, *Governing Without a Fundamental Direction of Time: Minimal Primitivism About Laws of Nature*, 2021. Preprint. https://arxiv.org/abs/2109.09226.

[15] R. Clifton, J. Bub, and H. Halvorson, *Characterizing Quantum Theory in Terms of Information-Theoretic Constraints*, Foundations of Physics, 33 (2003), pp. 1561–1591.

[16] A. Einstein, *What is the Theory of Relativity?*, London Times, 28 November (1919), pp. 53–54.

[17] A. Einstein, *Autobiographical notes*, in Albert Einstein: Philosopher-Scientist, P. Schilpp, ed., Open Court, La Salle, IL, 1949, pp. 3–94.

[18] A. Einstein, *Letter to Arnold Sommerfeld, January 14, 1908*, in The Collected Papers of Albert Einstein, Volume 5 The Swiss Years: Correspondence, 1902–14, M. J. Klein, A. J. Kox, J. Renn, and R. Schulman, eds., Princeton University Press, Princeton, NJ, 1995.

[19] P. Evans, *Retrocausality at No Extra Cost*, Synthese, 192 (2015), pp. 1139–1155.

[20] M. Giovanelli, *Relativity Theory as a Theory of Principles. A Reading of Cassirer's* Zur Einstein'schen Relativitätstheorie, HOPOS: The Journal of the International Society for the History of Philosophy of Science, 13 (2023), pp. 261–296.

[21] S. Goldberg, *Understanding Relativity: Origin and Impact of a Scientific Revolution*, Birkhäuser, Boston, MA, 1984.

[22] D. A. Howard and M. Giovanelli, *Einstein's philosophy of science*, in The Stanford Encyclopedia of Philosophy, E. N. Zalta, ed., Stanford University, 2019.

[23] R. Knight, *Physics for Scientists and Engineers with Modern Physics*, Pearson, San Francisco, CA, 2022.

[24] H. A. Lorentz, *The Theory of Electrons and Its Applications to the Phenomena of Light and Radiant Heat*, G. E. Stechert and Co., New York, 2nd ed., 1916.

[25] T. Maudlin, *Quantum Non-Locality and Relativity*, Wiley-Blackwell, Oxford, 2011.

[26] T. Maudlin, *Einstein didn't think time was an illusion*, iai News, 29 November (2022). https://iai.tv/articles/tim-maudlin-einstein-didnt-think-time-was-an-illusion-auid-2317?_auid=2020.

[27] J. Pearl, *Causality: Models, Reasoning, and Inference*, Cambridge University Press, New York, 2009.

[28] H. Price, *Causal Perspectivalism*, in Causation, Physics, and the Constitution of Reality: Russell's Republic Revisited, H. Price and R. Corry, eds., Oxford University Press, Oxford, 2007, pp. 250–292.

[29] H. Price and B. Weslake, *The Time Asymmetry of Causation*, in The Oxford Handbook of Causation, H. Beebee, C. Hitchcock, and P. Menzies, eds., Oxford University Press, New York, 2010, pp. 414–443.

[30] H. Price and K. Wharton, *Taming the quantum spooks*, Aeon, 14 September (2016). https://aeon.co/essays/can-retrocausality-solve-the-puzzle-of-action-at-a-distance.

[31] H. Price and K. Wharton, *Why entanglement?*, 2022. Preprint. https://arxiv.org/abs/2212.06986.

[32] G. Rubino, L. A. Rozema, A. Feix, M. Araújo, J. M. Zeuner, L. M. Procopio, C. Brukner, and P. Walther, *Experimental verification of an indefinite causal order*, Science Advances, 3 (2017), p. e1602589.

[33] B. Russell, *Mysticism and Logic: And Other Essays*, Longmans, Green and Co., New York, 1919.

[34] I. Salom, *2022 Nobel Prize in Physics and the End of Mechanistic Materialism*, Phlogiston, 31 (2023). In Press. https://arxiv.org/abs/2308.12297.

[35] R. Serway and J. Jewett, *Physics for Scientists and Engineers*, Brooks/Cole, Boston, MA, 10th ed., 2019.

[36] M. Silberstein, W. M. Stuckey, and T. McDevitt, *Beyond the Dynamical Universe: Unifying Block Universe Physics and Time as Experienced*, Oxford University Press, Oxford, 2018.

[37] M. Silberstein, W. M. Stuckey, and T. McDevitt, *Beyond Causal Explanation: Einstein's Principle not Reichenbach's*, Entropy, 23 (2021), p. 114.

[38] W. Van Camp, *Principle theories, constructive theories, and explanation in modern physics*, Studies in History and Philosophy of Science Part B: Studies in History and Philosophy of Modern Physics, 42 (2011), pp. 23–31.

[39] K. Wharton and R. Liu, *Entanglement and the Path Integral*, 2022. Preprint. https://arxiv.org/abs/2206.02945.

[40] K. Wharton, D. Miller, and H. Price, *Action duality: A constructive principle for quantum foundations*, Symmetry, 3 (2011), pp. 524–540.

[41] N. Wolchover, *Quantum Mischief Rewrites the Laws of Cause and Effect*, Quanta Magazine, 11 March (2021). https://www.quantamagazine.org/quantum-mischief-rewrites-the-laws-of-cause-and-effect-20210311/.

6

A Principle Response from Quantum Information Theorists

Can quantum theory be derived from simple principles, in a similar way as the Lorentz transformations can be derived from the relativity principle and the constancy of the speed of light? The exciting answer is "yes"!

Markus Müller Group, Institute for Quantum Optics and
Quantum Information (2021)

We have now laid out our case for pursuing a principle solution to the mystery of entanglement to resolve the EPR and EPR–Bell paradoxes and answer Mermin's challenge without violating locality, statistical independence, intersubjective agreement, or unique experimental outcomes and without altering quantum mechanics. As we will see, there are already principle accounts of quantum mechanics that do not have consensus support in the foundations community, so merely rendering quantum mechanics a principle theory is not a magic bullet leading to a consensus understanding of quantum mechanics.

All principle accounts of quantum mechanics are based on an empirically discovered fact, so what these extant accounts are all missing is a compelling fundamental principle to justify that empirically discovered fact. What we will show in this chapter is that quantum information theorists have based their principle account of quantum mechanics (called information-theoretic reconstructions) on an empirically discovered fact (Information Invariance & Continuity) that can be justified by the relativity principle, which has proven compelling to physicists for centuries. So, according to the information-theoretic reconstructions of quantum mechanics, the (finite-dimensional) Hilbert space formalism of quantum mechanics follows from NPRF + h, and since this includes the Bell states for the Mermin device of Chapter 2, NPRF + h provides a principle explanation of / solution to the mystery of entanglement (Chapter 7). Let's recap how we got here to see how we will proceed in this chapter.

Einstein's Entanglement. W. M. Stuckey, Michael Silberstein, and Timothy McDevitt, Oxford University Press.
© Oxford University Press (2024). DOI: 10.1093/9780198919698.003.0007

6.1 Recap and Direction

We started in Chapter 2 with an explanation of the mystery of entanglement for the general reader provided by Mermin. Using the Mermin device, we introduced the Bell inequality for a pair of spin-entangled particles and showed how that inequality is violated by certain measurements of a Bell spin triplet state (triplet state). The key piece of quantum formalism in that analysis was the probability that Alice and Bob obtain the same measurement outcomes in any particular trial of the experiment, both up (+1) or both down (–1), given that the angle between their Stern–Gerlach magnets was θ.

Quantum mechanics (represented by the Mermin device) says those joint probabilities are $P(++) = P(--) = \frac{1}{2}\cos^2(\theta/2)$. Accordingly, when $\theta = 0$ (case (a)), quantum mechanics says $P(++) = P(--) = \frac{1}{2}$, which means $P(++) + P(--) = 1$ so the probability of different outcomes is $P(+-) = P(-+) = 0$ (Fact 1). And, when $\theta = 120°$ (case (b)), quantum mechanics says $P(++) = P(--) = \frac{1}{8}$, which means $P(++) + P(--) = \frac{1}{4}$ (Fact 2). The Bell inequality resulting from a constructive explanation of case (a) assuming hidden variables, locality, and statistical independence says Alice and Bob must have agreement in at least $\frac{1}{3}$ of all case (b) trials (Bell inequality for Mermin device), while these joint probabilities for quantum mechanics predict the case (b) agreement to be $\frac{1}{4}$. That is, quantum mechanics violates the Bell inequality. Mermin concluded his paper with the following challenge:

> It is left as a challenging exercise to the physicist reader to translate the elementary quantum-mechanical reconciliation of cases (a) and (b) into terms meaningful to a general reader struggling with the dilemma raised by the device.

Since Mermin explicitly or tacitly adopted Bell's assumptions, he essentially ruled out constructive answers to his challenge (although, as we stated in Chapter 3, Mermin later opted to abandon intersubjective agreement). That is, as we outlined in Chapter 3, Bell's theorem shows how any constructive explanation of the violation of Bell's inequality entails the violation of locality, statistical independence, intersubjective agreement, and/or unique experimental outcomes.

In response to this state of affairs, quantum information theorist Chris Fuchs writes [19]:

What is the cause of this year-after-year sacrifice to the "great mystery?" Whatever it is, it cannot be for want of a self-ordained solution: Go to any meeting, and it is like being in a holy city in great tumult. You will find all the religions with all their priests pitted in holy war ... They all declare to see the light, the ultimate light. Each tells us that if we will accept their solution as our savior, then we too will see the light.

As Fuchs noted, the story from each "religion" is always, "Just give up [blank] and the mystery disappears!" Unfortunately for its advocates, everyone else considers [blank] essential to causal explanation. Maudlin agrees. Concerning what each "religion" says about the other, he writes [32]:

They may correctly note that according to every one of their rival theories, God was malicious, and having thus eliminated every other possibility, claim their own theory the victor. The problem is that *every* partisan can argue in this way since *every* theory posits some funny business on the part of the Deity.

In Chapter 4, we likened the situation today concerning the mystery of entanglement to the situation in the late 19th century concerning the mystery of length contraction. Einstein's response there was to abandon "constructive efforts" and propose a principle solution to that mystery, i.e., the relativity principle justifies the light postulate giving length contraction.

In Chapter 5, we argued that causal accounts of "the elementary quantum-mechanical reconciliation of cases (a) and (b)" could be salvaged as long as the asymmetrical causality of classical mechanics was replaced with the symmetrical co-causality of quantum mechanics. This notion of quantum co-causality is consistent with atemporal or all-at-once explanation of Bell state correlations in accord with an appropriately modified Reichenbach's Principle and is gaining interest in the foundations community. Of course, quantum co-causality still violates locality and/or statistical independence (depending on the co-causal mechanism), so in order to solve the mystery of entanglement without violating locality and/or statistical independence, we are motivated to seek an

atemporal / all-at-once / AGC principle solution to the mystery of entanglement, and quantum information theorists have been involved in precisely this
approach for many years.

6.2 Brief History of Information-Theoretic Reconstructions

For example, Fuchs wrote [20, p. 285]:

> Associated with each system [in quantum mechanics] is a complex vector
> space. Vectors, tensor products, all of these things. Compare that to one of
> our other great physical theories, special relativity. One could make the state-
> ment of it in terms of some very crisp and clear physical principles: The
> speed of light is constant in all inertial frames, and the laws of physics are
> the same in all inertial frames. And it struck me that if we couldn't take the
> structure of quantum theory and change it from this very overt mathemati-
> cal speak—something that didn't look to have much physical content at all,
> in a way that anyone could identify with some kind of physical principle—if
> we couldn't turn that into something like this, then the debate would go on
> forever and ever. And it seemed like a worthwhile exercise to try to reduce
> the mathematical structure of quantum mechanics to some crisp physical
> statements.

And Lucien Hardy wrote [25, p. 224]:

> The standard axioms of [quantum mechanics] are rather ad hoc. Where does
> this structure come from? Can we write down natural axioms, principles,
> laws, or postulates from which we can derive this structure? Compare with
> the Lorentz transformations and Einstein's two postulates for special relativ-
> ity. Or compare with Kepler's Laws and Newton's Laws. The standard axioms
> of quantum theory look rather ad hoc like the Lorentz transformations or
> Kepler's laws. Can we find a natural set of postulates for quantum theory that
> are akin to Einstein's or Newton's laws?

So, in 2000 "Christopher Fuchs implored the community to 'find an information theoretic reason' for the axioms of [quantum mechanics]" [25]. The
first to respond to that challenge was Hardy with his 2001 paper [24] "Quantum theory from five reasonable axioms." Since then, many quantum information theorists have published so-called axiomatic reconstructions of quantum
mechanics based on information-theoretic principles. The idea is based on

the assumption that quantum mechanics is a probability theory about information. As Zeilinger put it [15]:

> The distinction between reality and our knowledge of reality, between reality and information, cannot be made. There is no way to refer to reality without using the information we have about it.

While this is the starting point for information-theoretic reconstructions specifically, there is a history predating Fuchs' challenge that we should summarize.

For example, in 1996 Carlo Rovelli wrote [36]:

> [Q]uantum mechanics will cease to look puzzling only when we will be able to *derive* the formalism of the theory from a set of simple physical assertions ("postulates," "principles") about the world. Therefore, we should not try to *append* a reasonable interpretation to the quantum mechanics *formalism*, but rather to derive the formalism from a set of experimentally motivated postulates.

Notice he is advocating a principle approach to quantum mechanics, since "experimentally motivated postulates" may certainly be replaced with "empirically discovered facts." And, he is motivated by the success of special relativity specifically [36]:

> The reasons for exploring such a strategy are illuminated by an obvious historical precedent: special relativity. ... Special relativity is a well understood physical theory, appropriately credited to Einstein's 1905 celebrated paper. The formal content of special relativity, however, is coded into the Lorentz transformations, written by Lorentz, not by Einstein, and before 1905. So, what was Einstein's contribution? It was to understand the physical meaning of the Lorentz transformations.

While Rovelli did not explicitly produce a complete derivation of the quantum-mechanical formalism in his 1996 paper, Alexei Grinbaum did so in 2003–2005 starting with the two information-theoretic axioms in Rovelli's 1996 paper [21, 22] (Section 6.6).

Likewise, Zeilinger introduced his Foundational Principle for quantum mechanics in 1999, writing [43]:

> Physics in the 20th century is signified by the invention of the theories of special and general relativity and of quantum theory. Of these, both the

special and the general theory of relativity are based on firm foundational principles, while quantum mechanics lacks such a principle to this day. By such a principle, I do not mean an axiomatic formalization of the mathematical foundations of quantum mechanics, but a foundational conceptual principle. In the case of the special theory, it is the Principle of Relativity, ... In the case of the theory of general relativity, we have the Principle of Equivalence ... Both foundational principles are very simple and intuitively clear ...

I submit that it is because of the very existence of these fundamental principles and their general acceptance in the physics community that, at present, we do not have a significant debate on the interpretation of the theories of relativity. Indeed, the implications of relativity theory for our basic notions of space and time are broadly accepted.

In his 1999 paper "Quantum Mechanics as a Principle Theory," Jeffrey Bub writes [6]:

I show how quantum mechanics, like the theory of relativity, can be understood as a "principle theory" in Einstein's sense ...

The idea that quantum mechanics deals fundamentally with information goes back at least to Neils Bohr, as summed up by this famous 1958 statement by Aage Petersen about Bohr's belief [33]:

There is no quantum world. There is only an abstract quantum physical description. It is wrong to think that the task of physics is to find out how nature is. Physics concerns what we can say about Nature.

As Carlton Caves, Chris Fuchs, and Rüdiger Schack note [10], this is particularly relevant "for information-based interpretations of quantum mechanics, where quantum states, like probabilities, are taken to be states of knowledge rather than states of nature." And Wheeler, who was greatly influenced by Bohr, wrote in 1989 [42]:

No element in the description of physics shows itself as closer to primordial than the elementary quantum phenomenon, that is, the elementary device-intermediated act of posing a yes–no physical question and eliciting an answer or, in brief, the elementary act of observer-participancy. Otherwise stated, every physical quantity, every it, derives its ultimate significance from bits, binary yes-or-no indications, a conclusion which we epitomize in the phrase, it from bit.

This is Wheeler's famous "It from Bit" hypothesis, i.e., that physical objects (It) are based on information (Bit). Of course, quantum mechanics might be dealing with information most fundamentally even if the converse of Wheeler's hypothesis is true, i.e., information requires physical objects to exist. For example, John Preskill writes [35]:

> The moral we draw is that "information is physical" and it is instructive to consider what physics has to tell us about information. But fundamentally, the universe is quantum mechanical. How does quantum theory shed light on the nature of information?

Rolf Landauer agrees, stating [30]:

> Information is not a disembodied abstract entity; it is always tied to a physical representation.

Luckily, we do not need to commit either way on the ontology here, both extremes are compatible with a principle account of quantum mechanics. All that matters to us is that quantum mechanics can be understood as a theory about information. That fact, and a particular concern of Feynman, motivated Zeilinger to posit his Foundational Principle for quantum mechanics in 1999 [43]. Feynman's concern was voiced in this 1982 statement [18, p. 57]:

> It always bothers me that, according to the laws as we understand them today, it takes a computing machine an infinite number of logical operations to figure out what goes on in no matter how tiny a region of space and no matter how tiny a region of time ... why should it take an infinite amount of logic to figure out what one tiny piece of space-time is going to do?

Bub (later with Rob Clifton and Hans Halvorson) focused on the algebraic difference between classical (Boolean algebra) and quantum (non-Boolean algebra) possibility spaces per Heisenberg's commutative (classical mechanics) versus noncommutative (quantum mechanics) algebra of observables. As we pointed out at the end of Chapter 5, Bub compared this difference with the difference between the geometry of Euclidean spacetime (Newtonian mechanics) and Minkowski spacetime (special relativity) [7, 8, 9]. Rovelli also focused on this difference in commutativity by noting that information gained in measurement about some property of a quantum system is lost when subsequently measuring a noncommutative/non-Boolean complementary property of that system [36] (Section 3.2). Here, we will follow Hardy's approach with its important precursors, and that returns us to Zeilinger's Foundational Principle as a response to Feynman's concern.

Zeilinger actually wrote two different forms of his Foundational Principle as given in this single statement by Gregg Jaeger [28]:

"An elementary system carries 1 bit of information," because "an elementary system represents the truth value of one proposition."

In the abstract of Zeilinger's 1999 paper, we see how his Foundational Principle leads to the mystery of entanglement [43]:

In contrast to the theories of relativity, quantum mechanics is not yet based on a generally accepted conceptual foundation. It is proposed here that the missing principle may be identified through the observation that all knowledge in physics has to be expressed in propositions and that therefore the most elementary system represents the truth value of one proposition, i.e., it carries just one bit of information. Therefore an elementary system can only give a definite result in one specific measurement. The irreducible randomness in other measurements is then a necessary consequence. For composite systems entanglement results if all possible information is exhausted in specifying joint properties of the constituents.

This was expanded by Brukner and Zeilinger in 1999 [3]:

The total information carried by the system is invariant under such transformation from one complete set of complementary variables to another.

They further clarified this in a 2003 paper where they wrote [4]:

We show that if, in our description of Nature, we use one definite proposition per elementary constituent of Nature, some of the essential characteristics of quantum physics, such as the irreducible randomness of individual events, quantum complementary and quantum entanglement, arise in a natural way. Then quantum physics is an elementary theory of information.

And with this statement in that same 2003 paper:

Thus, if we gradually change the orientation of the magnets in a set of Stern–Gerlach apparata defining a complete set of mutually complementary observables, a continuous change of the information vector will result.

(again, as in Section 3.2) everything is in place for their 2009 statement of Information Invariance & Continuity [5]:

The total information of one bit is invariant under a continuous change between different complete sets of mutually complementary measurements.

Borivoje Dakić and Brukner emphasize the importance of continuity in the third axiom of their 2009 reconstruction [13]:

> (3) (Reversibility) Between any two pure states there exists a reversible transformation. If one requires the transformation from the last axiom to be continuous, one separates quantum theory from the classical probabilistic one.

Dakić and Brukner's reconstruction greatly influenced one of the "first fully rigorous, complete reconstructions" [34] in 2010[1] by Lluís Masanes and Markus Müller. Concerning continuity they write [31]:

> ... if Requirement 4 is strengthened by imposing continuity of the reversible transformations, then [classical probability theory] is ruled out and [quantum theory] is the only theory satisfying the requirements. This strengthening can be justified by the continuity of time evolution of physical systems.

As we will see in Section 6.4, their wording here is similar to that in Hardy's 2001 reconstruction, where we see the same emphasis on continuity. Finally, Grinbaum writes [23]:

> We see that various axiomatic systems for quantum theory contain, under one form or another, the assumption of continuity, and it is this assumption which is largely responsible for making things quantum.

There is one more requirement needed to reproduce quantum probability theory: a way to stipulate how the rest of the probability space (or possibility/Hilbert space) can be constructed from the qubit (we will provide an example in Subchapter 9C). This is the second axiom for Dakić and Brukner:

> (2) (Locality) The state of a composite system is completely determined by local measurements on its subsystems and their correlations.

This simply means there is nothing "hidden" or "missing" from the quantum formalism [2]. It is akin to having to assume linearity in addition to the light postulate in order to derive the Lorentz transformations.

---

[1] The other is due to Giulio Chiribella, Giacomo D'Ariano, and Paolo Perinotti in 2010 [11].

6.3 What's Missing?

These principle approaches have clearly not won consensus support in the foundations community or we would not be writing this book. One problem with them according to Dakić and Brukner is their "highly abstract mathematical assumptions without an immediate physical meaning" [13]. For example, here are the "five simple physical requirements" of Masanes and Müller's reconstruction cited above:

1. In systems that carry one bit of information, each state is characterized by a finite set of outcome probabilities.
2. The state of a composite system is characterized by the statistics of measurements on the individual components.
3. All systems that effectively carry the same amount of information have equivalent state spaces.
4. Any pure state of a system can be reversibly transformed into any other.
5. In systems that carry one bit of information, all mathematically well-defined measurements are allowed by the theory.

These establish classical probability theory and quantum probability theory uniquely among all generalized probabilistic theories. As stated above, if you change Requirement 4 to read, "Any pure state of a system can be *continuously* reversibly transformed into any other," then you select quantum probability theory alone. This is the kinematics of quantum mechanics; to get the dynamics (measurement update and Schrödinger's equation) you can add two more requirements [31]:

- If a system is measured twice "in rapid succession" with the same measurement, the same outcome is obtained both times.
- Closed systems evolve reversibly and continuously in time.

Masanes and Müller's reconstruction is essentially Hardy's reconstruction (Section 6.4) minus his (unnecessary) Simplicity axiom [34]. You can see the problem—unless you possess a 'physical intuition' for generalized probabilistic theories, these are probably not going to convey much about physical reality to you.

Besides a lack of "immediate physical meaning," another common complaint from the foundations community about the reconstruction program is noted here by Van Camp [39]:

However, nothing additional has been shown to be incorporated into an information-theoretic reformulation of quantum mechanics beyond what is contained in quantum mechanics itself. It is hard to see how it could offer more unification of the phenomena than quantum mechanics already does since they are equivalent, and so it is not offering any explanatory value on this front.

The unification that Van Camp alludes to here can be seen in quantum mechanics as a structural explanation per Felline [16, 17]. For example, the Heisenberg uncertainty principle (Chapter 3) in its most general form applies to any pair of complementary variables, such as position x and momentum p or spin S_x and spin S_z. As Felline points out [16]:

> There is in fact no apparent sense in which the processes underlying the loss of a determinate position for a particle with definite momentum can be said to be the same as the one leading to the loss of spin S_x for a particle with determinate spin S_z. Such a relation is instead explained as part and parcel of the algebraic structure of observables in Quantum Theory.

Van Camp's complaint about the reconstruction program is that it does not provide more unification than we already have in standard quantum mechanics.

In short, while many in foundations are satisfied with structural explanation per the formalism of special relativity (spacetime geometry per the Lorentz transformations), few are satisfied with structural explanation per the formalism of quantum mechanics (finite-dimensional Hilbert space), and the reconstruction program does not provide any help here. Indeed, for those without a 'physical intuition' for generalized probabilistic theories, the reconstruction program obfuscates, rather than clarifies, structural explanation per quantum mechanics. That is why we propose completing the reconstruction program via principle explanation.

What we show in the remainder of this chapter is that the principle of Information Invariance & Continuity residing at the foundation of information-theoretic reconstructions (one way or another) represents an empirically discovered fact, i.e., the invariance of Planck's constant h between different inertial reference frames. The direct mathematical consequence of this "Planck postulate" is noncommutativity/superposition/complementarity, i.e., the 'weirdness' of the qubit upon which the kinematic structure of quantum mechanics is built in the reconstruction program.

So, we see immediately what the reconstruction program is missing, i.e., a compelling fundamental principle justifying Information Invariance & Continuity (Planck postulate). Foundations in general is not going to accept that the noncommutativity/superposition/complementarity of the qubit solves the mystery of entanglement any more than they would accept the light postulate as a solution to the mystery of length contraction. The reason people can accept that the mystery of length contraction is solved by the light postulate is because the light postulate is a necessary consequence of the relativity principle and the relativity principle is compelling enough to provide a foundational explanans.

That is why Höhn's comment in Chapter 2 that [27] "Entanglement from complementarity is not as intuitive as the relativity of simultaneity from the relativity principle" misses the analogy completely. The proper analogy, as we show in this chapter and the next, is that *both* the relativity of simultaneity *and* entanglement follow most fundamentally from the relativity principle. Likewise, the quote from Müller's group at IQOQI at the beginning of this chapter misses the analogy and fails to capture the true accomplishment of the reconstruction program. A corrected version of the quote would read:

> Can quantum theory be derived from simple principles, in a similar way as the Lorentz transformations can be derived from ~~the relativity principle and~~ the constancy of the speed of light? The exciting answer is "yes"!

In other words, the reconstruction program has succeeded in rendering quantum mechanics a principle theory with a corresponding structural explanation, i.e., a mathematical formalism (Hilbert space of quantum mechanics) derived from an empirically discovered fact (Information Invariance & Continuity). What they do not have, as indicated by the struck-through text in the corrected quote, is quantum mechanics as a principle *explanation*. Again, the difference between structural explanation per a principle theory and a principle explanation (Chapter 1) is that the explanans for a structural explanation is the formalism of the principle theory (e.g., geometry of spacetime per the Lorentz transformations in special relativity), while the explanans for a principle explanation is a compelling fundamental principle justifying the empirically discovered fact at the foundation of the principle theory (e.g., the relativity principle justifying the light postulate).

Alternatively, one could propose a constructive account of the empirically discovered fact at the foundation of the principle theory, e.g., length contraction due to movement through the aether causing the light postulate. No consensus aether account has ever been rendered for the light postulate, while its justification via the relativity principle is widely accepted. Likewise, the reconstruction program has derived the Hilbert space of quantum mechanics from Information Invariance & Continuity, but has not provided a constructive account of Information Invariance & Continuity or a compelling fundamental principle to justify it. Essentially, the constructive accounts of violations of Bell's inequality as outlined in Chapter 3 are the 'aether theories' for Information Invariance & Continuity in this analogy and Bell's theorem tells us why they have not received (nor will they ever likely receive) consensus support.

Consequently, again, we propose a completion of the reconstruction program via principle explanation [38] according to Fuchs, Hardy, and Rovelli's desideratum quoted at the outset of Section 6.2. It is obvious that the Planck postulate can be justified by the relativity principle, so what we will do in this chapter is explain how Information Invariance & Continuity entails the Planck postulate. To do that, we start with an overview of Hardy's original approach to the reconstruction program.

6.4 Hardy's Reconstruction

As Hardy notes in his first reconstruction paper [24], if you delete just one word ("continuous") from his Axiom 5, "then we obtain classical probability theory instead" of quantum probability theory. [Again, in Masanes and Müller's reconstruction you add the word "continuously" to their Requirement 4 to select quantum probability theory.] We explained what that meant in Chapter 0 and we will review it again shortly, but first let's look at what sets Hardy's approach apart from most textbook presentations of quantum mechanics.

What Hardy did was to cast quantum mechanics as a probability theory in the manner of classical probability theory. George Mackey and Günther Ludwig did likewise in the 1950s using quantum logic [28], but Hardy dropped the quantum logic approach. Most accounts of quantum mechanics are in terms of the wavefunction (quantum state), which is not a probability. The wavefunction is a probability amplitude that evolves in time in a possibility space (Hilbert space) and it must be squared to get a probability. In classical probability theory, you deal directly with probabilities, so by casting quantum mechanics as a theory of probabilities Hardy was able to compare it directly with classical probability theory. Now let's look again at what Hardy meant when he said continuity is the one word that distinguishes quantum probability theory from classical probability theory.

As we stated in Chapter 2, both theories are built upon a fundamental piece of information, i.e., the binary answer (yes–no, up–down, etc.) to some question. In classical probability theory, this fundamental unit of information is called a *classical bit*. In quantum probability theory, this fundamental unit of information is called a quantum bit or *qubit* for short.

Again, the information-theoretic reconstructions of the (finite-dimensional) Hilbert space formalism for quantum mechanics are based on the qubit in general. While this is clearly a strength of information-theoretic reconstructions, the general reader will probably find such a general approach less tangible than specific physical instantiations (recall Masanes and Müller's example above). Plus, without a clear connection to empirical reality the reconstructions are difficult to relate to physical principles. That is why we have restricted our explanations to actual physical examples like the spin-$\frac{1}{2}$ particle of Chapter 2. Again, as Brukner and Zeilinger said [3], "spin-$\frac{1}{2}$ affords a model of the quantum mechanics of all two-state systems, i.e., qubits" (see also Masanes and Müller's Requirement 3 above). We will add photon polarization in Chapter 7 and the double-slit experiment in Subchapter 9C.

The model of a classical bit here will be a ball and a box. Suppose we have a box that may contain a ball or not. There is only one measurement we can make: we can open the box. The measurement outcome is a yes–no answer to the question, "Does this box contain a ball?" Since the ball either resides in the box or it doesn't, the probability that it resides in the box (p_1) plus the probability that it doesn't reside in the box (p_2) must sum to 1, i.e., $p_1 + p_2 = 1$. We can depict that fact with a vector $\vec{p}$ in a vector space with axes $|1\rangle$ and $|2\rangle$ (Figure 6.1). $|1\rangle$ is the "yes" answer to the measurement and $|2\rangle$ is the "no" answer to the measurement.

The vectors $|1\rangle$ and $|2\rangle$ are called *pure states*, as they correspond to actual single measurement outcomes. Since there is only one measurement possible on this classical bit, the state vector $\vec{p}$ pointing along the line in Figure 6.1 does not represent a possible single measurement outcome. It represents what is called a *mixed state*, i.e., a distribution of outcomes for that single measurement. Notice that the pure states for the classical bit are not connected via other pure states, they are only connected by mixed states. Now let's look at how this differs from the qubit.

As we said, we will model our qubit with spin-$\frac{1}{2}$ (or spin for short), which is what we used to physically instantiate the Mermin device in Chapter 2.

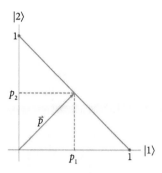

Fig. 6.1 Probability state space for the classical bit.

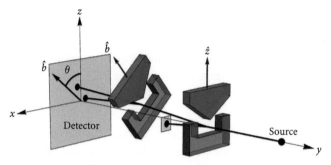

Fig. 6.2 In this setup, the first SG magnets (oriented at $\hat{z}$) are being used to produce an initial spin wavefunction $|\psi\rangle = |z+\rangle$ for measurement by the second SG magnets (oriented at $\hat{b}$). Compare this with Figure 6.3.

Suppose we prepare our particle in the spin wavefunction $|\psi\rangle = |z+\rangle$ and make SG measurements along $\hat{b}$ making an angle θ with $\hat{z}$, as shown in Figure 6.2. Quantum mechanics tells us that the probability p of getting a +1 outcome when measuring the spin along $\hat{b}$ is $p(+ \mid \theta) = \cos^2(\theta/2)$. So, when $\theta = 0$ ($\hat{b} = \hat{z}$) we get $p = 1$, when $\theta = 90°$ we get $p = \frac{1}{2}$, and when $\theta = 180°$ we get $p = 0$, as shown in Figure 6.3. Our qubit probability space is a sphere called the *Bloch sphere* (Figure 6.3). Since this state space is equivalent to three-dimensional real space, the Bloch sphere is shown in a real space reference frame.

Notice that every vector from the origin to the surface of the Bloch sphere corresponds to the +1 outcome of a single SG spin measurement in the direction $\hat{b}$ shown, just as the vector $|1\rangle$ corresponds to the "yes" outcome of our box measurement. So, every vector on the Bloch sphere is a pure state (the mixed states are inside the Bloch sphere, but we are not concerned with them). Now

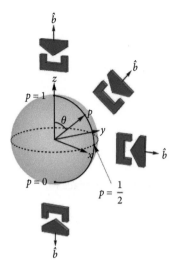

Fig. 6.3 Probability space and real space for the qubit associated with the spin wavefunction $|\psi\rangle = |z+\rangle$. Compare this with Figure 6.2.

we can see the difference that Hardy pointed out between classical and quantum probability theories—the pure states for the classical bit are distributed discretely in their probability space while the pure states for the qubit are distributed *continuously* in their probability space.

6.5 Brukner and Zeilinger's Extrapolation

Those of you familiar with quantum mechanics will realize this is just another way to describe superposition, as we explained in Chapter 0. In quantum mechanics, if the wavefunction $|z+\rangle$ represents the probability amplitude with the +1 outcome ("yes" answer) for an SG spin measurement along $\hat{z}$ and the wavefunction $|z-\rangle$ represents the probability amplitude with the −1 outcome ("no" answer) for an SG spin measurement along $\hat{z}$, then these wavefunctions are perpendicular vectors in Hilbert space, a vector space where wavefunctions live (Figure 6.4). According to quantum mechanics, any linear combination of $|z+\rangle$ and $|z-\rangle$ represents the probability amplitude with the +1 outcome ("yes" answer) for some other SG spin measurement (superposition), and its corresponding probability amplitude with the −1 outcome ("no" answer) is perpendicular to it.

For example, for SG spin measurements in the xz plane $(|z+\rangle + |z-\rangle)/\sqrt{2} = |x+\rangle$, the probability amplitude with the +1 outcome ("yes" answer) to an SG spin measurement along $\hat{x}$ (Figure 6.4). An orthogonal pair of Hilbert space vectors corresponds to opposing probability vectors in Figure 6.3, e.g., $\theta = 0$ and $\theta = 180°$ correspond to $|z+\rangle$ and $|z-\rangle$, respectively. So, superposition leads to the continuous distribution of pure states for the Bloch sphere.

Notice also that we can identify any particular $\hat{b}$ with a set of three mutually orthogonal directions in real space and relate those directions to their corresponding SG spin measurements (Figure 6.5). These reference frames are related by spatial rotations, which are part of Galilean and Lorentz transformations between inertial reference frames. This is what Brukner and Zeilinger meant by "different complete sets of mutually complementary measurements" in their principle of Information Invariance & Continuity.

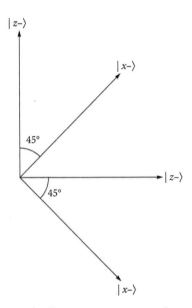

Fig. 6.4 The basis vectors of Hilbert space represent the possible outcomes of a measurement. Here we see the basis vectors for an SG spin measurement along $\hat{z}$ ($|z+\rangle$ representing a +1 outcome and $|z-\rangle$ representing a –1 outcome) and the basis vectors for an SG spin measurement along $\hat{x}$. If $|\psi\rangle = |z+\rangle$, as shown in Figure 6.2, and $\hat{b} = \hat{x}$, then we project $|\psi\rangle = |z+\rangle$ onto $|x+\rangle$ and square the result $(1/\sqrt{2})$ to find the probability $(\frac{1}{2})$ that the outcome of our SG measurement of $|\psi\rangle = |z+\rangle$ along $\hat{x}$ will produce a +1 outcome. Likewise for a –1 outcome.

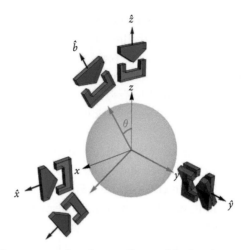

Fig. 6.5 Probability space and real space for a qubit showing two reference frames of mutually complementary SG spin measurements.

That the fundamental unit of information (qubit) lives in a probability space with the same dimension as real space is justified by Dakić and Brukner using their *closeness requirement* [13]:

> the dynamics of a single elementary system can be generated by the invariant interaction between the system and a "macroscopic transformation device" that is itself described within the theory in the macroscopic (classical) limit.

That makes sense if you understand that the quantum systems being measured need to make correspondence with their classical measuring devices. For example, the classical magnetic field of an SG magnet is used to measure the spin of spin-$\frac{1}{2}$ particles, and that classical magnetic field "can be seen as a limit of a large coherent state, where a large number of spin-$\frac{1}{2}$ particles are all prepared in the same quantum state" [13].

The information contained in the qubit of Figure 6.3 is simply "+1 is the outcome of an SG spin measurement in the $\hat{z}$ direction." Now suppose you ask, "Is +1 the outcome of an SG spin measurement in the $\hat{b}$ direction?" The answer is indefinite, i.e., probabilistic, as we explained in Section 3.2 when we showed how "the amount of information contained by a qubit for some binary-outcome measurement is fixed (invariant) and distributed according to the Pythagorean theorem between the binary outcomes of different measurements on that qubit." This is what Zeilinger meant when he wrote [43], "an elementary system can only give a definite result in one specific measurement.

The irreducible randomness in other measurements is then a necessary consequence." So, we see that the probabilistic character of quantum mechanics follows from Zeilinger's Foundational Principle.

Finally, we point out that one always obtains ±1 (short for ±$\hbar$/2, as described in Chapter 2) for an SG spin measurement at any $\hat{b}$. That is what Brukner and Zeilinger meant by "The total information of one bit is invariant" (Section 3.2). If the fundamental unit of information was not "invariant under a continuous change between different complete sets of mutually complementary measurements," then we would be getting a fraction of ±1 at different $\hat{b}$. Since, as Weinberg pointed out [40], measuring an electron's spin via SG magnets constitutes the measurement of "a universal constant of nature, Planck's constant h," we see that Information Invariance & Continuity entails the invariant measurement of Planck's constant h (Planck postulate). But, always obtaining ±1 rather than fractional outcomes is counterintuitive.

Thinking of spin angular momentum as a vector in real space, it would make a lot more sense if our SG spin measurement of $|z+\rangle$ along $\hat{b}$ produced the projection of $+1\hat{z}$ along $\hat{b}$, i.e., cos (θ), as shown in Figure 6.6. But if that happened, we would only be obtaining h for our measurement outcome in the $\hat{z}$ reference frame while all other inertial reference frames would be measuring a fraction of h. This is totally analogous to measuring different values of c when moving at different speeds relative to the light source. Both are violations of NPRF.

Adherence to the relativity principle for the measurement of c (light postulate) leads to time dilation and length contraction as we showed in Chapter

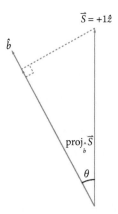

$$\vec{S} = +1\hat{z}$$

$\hat{b}$

$\text{proj}_{\hat{b}}\vec{S}$

θ

Fig. 6.6 The spin angular momentum of Bob's particle $\vec{S}$ projected along his measurement direction $\hat{b}$. This does *not* happen with spin angular momentum due to NPRF.

4. Adherence to NPRF for the measurement of h (Planck postulate) leads to 'average-only' projection. That is, while an SG measurement cannot produce an outcome of $\cos(\theta)$ at $\hat{b}$ because of NPRF, our ±1 results can *average* to the expected $\cos(\theta)$. Indeed, this is exactly what quantum mechanics gives us. In a 2022 paper, Olivier Darrigol justifies 'average-only' projection [14]:

> This is so because by a correspondence argument we expect the total angular momentum (or magnetic moment) of a large number of identically prepared, non-interacting spin-particles to behave as the angular momentum of a macroscopic object under measurement.

This is also justified by Dakić and Brukner's closeness requirement and Claude Comte's "homogeneity of statistical ensembles" [12].

Recall, to get the probability we have to square the probability amplitude. Specifically, we project $|\psi\rangle$ onto $|b+\rangle$ and square to obtain the probability that an SG spin measurement of $|\psi\rangle = |z+\rangle$ along $\hat{b}$ will produce $+1$. Likewise, we project $|\psi\rangle$ onto $|b-\rangle$ and square to obtain the probability that an SG spin measurement of $|\psi\rangle$ along $\hat{b}$ will produce -1.

From Figures 6.3 and 6.4 we see that the angle between $\hat{z}$ (for $|\psi\rangle = |z+\rangle$) and $\hat{b}$ (for $|b+\rangle$) is θ while the angle between $|z+\rangle$ and $|b+\rangle$ in Hilbert space is $\theta/2$. So, we have $P(b+ \mid \theta) = \cos^2(\theta/2)$ and $P(b- \mid \theta) = \sin^2(\theta/2)$, giving an average (expectation value) of

$$(+1)\cos^2\left(\frac{\theta}{2}\right) + (-1)\sin^2\left(\frac{\theta}{2}\right) = \cos(\theta).$$

In fact, one could use NPRF + h and the closeness requirement to demand

$$(+1)P(b+ \mid \theta) + (-1)P(b- \mid \theta) = \cos(\theta).$$

With that equation and the normalization

$$P(b+ \mid \theta) + P(b- \mid \theta) = 1$$

we can then *derive* the quantum-mechanical probabilities for our qubit.

So, Information Invariance & Continuity provides us with an empirically discovered fact (Planck postulate) that can be justified by the relativity principle (NPRF + h) in complete analogy with the relativity principle justifying the light postulate (NPRF + c). NPRF + h leads to the mystery of 'average-only' projection (superposition for spin-$\frac{1}{2}$) while NPRF + c leads to the mystery of length contraction. Consequently, Adam Koberinski and Markus Müller write [29]:

We suggest that (continuous) reversibility may be the postulate which comes closest to being a candidate for a glimpse on the genuinely physical kernel of "quantum reality." Even though Fuchs may want to set a higher threshold for a "glimpse of quantum reality," this postulate is quite surprising from the point of view of classical physics: when we have a discrete system that can be in a finite number of perfectly distinguishable alternatives, then one would classically expect that reversible evolution must be discrete too. For example, a single bit can only ever be flipped, which is a discrete indivisible operation. Not so in quantum theory: the state $|0\rangle$ of a qubit can be continuously-reversibly "moved over" to the state $|1\rangle$. For people without knowledge of quantum theory (but of classical information theory), this may appear as surprising or "paradoxical" as Einstein's light postulate sounds to people without knowledge of relativity.

6.6 Relational Quantum Mechanics as a Reconstruction

As we stated above, Grinbaum was able to render relational quantum mechanics an information-theoretic reconstruction based on Rovelli's following two information-theoretic axioms [23]:

1. There exists a maximum amount of relevant information that can be extracted from a system.
2. It is always possible to obtain new information about the system.

These axioms are the information-theoretic counterpart to what we quoted in Chapter 3, i.e., "relational quantum mechanics is an interpretation of quantum mechanics based on the idea that quantum states describe not an absolute property of a system but rather a relationship between systems" [1].

For example, Axiom 1 can be mapped to the fact that the qubit gives us a definite "yes" or "no" answer to only one question, e.g., "Will the spin measurement along $\hat{z}$ produce a +1 outcome?" Axiom 2 can be mapped to the fact that after we do a spin measurement of that qubit along $\hat{b} \neq \hat{z}$, the qubit now gives us a definite "yes" or "no" answer to only this new question, e.g., "Will the spin measurement along $\hat{b}$ produce a +1 outcome?" (Section 3.2). You can see how this relates to Brukner and Zeilinger's "Information Invariance," but we still need "Continuity," so where is continuity in this reconstruction?

In addition to Axioms 1 and 2, Grinbaum needed "the help of a few quantum logical assumptions," e.g., "the lattice of questions being isomorphic to the lattice of all closed subspaces of a Banach space constructed over a numeric field (i.e. real or complex numbers or quaternions)" [23]. This is where continuity is brought into the reconstruction.

As we have pointed out and as Grinbaum notes, reconstructions are "devoid of ontological commitments," which is deemed necessary for an interpretation. However, Grinbaum points out that:

> reconstruction appears more appealing than a mere interpretation as it leaves room for any justification of first principles, some such justifications being possibly different from ours. Indeed, one may equally well choose to adopt a specific ontological picture to justify [Axioms 1 and 2]. At the same time, regardless of a concrete philosophical justification for first principles, the meaning of quantum theory stands clear: it is a general theory of information constrained by several information-theoretic principles.

Regardless, as we noted in Chapters 3 and 5, Rovelli has also articulated an ontology (all-at-once or AGC), so relational quantum mechanics is both a reconstruction and interpretation at this point. Therefore, given Rovelli's statements above regarding special relativity as a precedent, and given that his proposed axioms in that 1996 paper were eventually used to satisfy his desideratum (Grinbaum's reconstruction and NPRF + h, as explained in this book), we might say that Rovelli is the progenitor for this new understanding of quantum mechanics in 1996.

6.7 Summary

To summarize, a fundamental principle for the axiomatic reconstructions of quantum mechanics using information-theoretic principles is Information Invariance & Continuity. According to this principle, there exists a fundamental unit of information (the qubit) that represents a definite outcome with respect to only one measurement. Therefore, other measurements of this qubit produce indefinite/probabilistic results. That is why quantum mechanics must be a probability theory.

Even though the results of other measurements are indefinite, the total information of the qubit is invariant (Section 3.2). Each measurement is associated with an inertial reference frame per its complementary measurements, and these inertial reference frames are related by spatial rotations (or spatial translations, see the double-slit experiment in Subchapter 9C). These facts tell us that the total information of the most fundamental unit of information is the same in all inertial reference frames, regardless of their relative spatial orientation (or location) in accord with the relativity principle, i.e., NPRF. All of this can be associated directly with Planck's constant h.

Recall from Chapter 3 that Planck's constant h represents "a universal limit on how much simultaneous information is accessible to an observer" [26]. The measure of available simultaneous information is given mathematically by the commutator for a complete set of mutually complementary measurements, and those complementary measurements establish a reference frame.

For example, for the complementary spin measurements (S_x, S_y, S_z) the commutator is $S_x S_y - S_y S_x = i\hbar S_z$, as given in Chapter 3. If $h = 0$, then $\hbar = h/2\pi = 0$ and those spin measurements would commute. But, if they commuted, we would have the classical situation instead of the quantum situation. [Again, in general, we obtain classical results from quantum mechanics when $h \to 0$.] That means the measurement of $|z+\rangle$ along $\hat{x}$ or $\hat{y}$ *would* produce the projection of $+1\hat{z}$ along $\hat{x}$ or $\hat{y}$, as shown in Figure 6.6, which in this case would be zero $(\cos(90°) = 0)$. In other words, given that we know the outcome of a $\hat{z}$ measurement is definitely going to be +1, we would also know the $\hat{x}$ and $\hat{y}$ measurement outcomes will be zero. But, if that happened, our measurements would only be producing h in the $\hat{z}$ direction while measurements in all other directions would be producing a fraction of h in violation of NPRF.

This is totally analogous to measuring the speed of light relative to some source and obtaining a value of zero. The understanding would be that the measuring device is at rest with respect to the light beam. This violates the light postulate and NPRF.

Putting this all together we see that Information Invariance & Continuity at the foundation of axiomatic reconstructions of quantum mechanics is the information-theoretic counterpart to the conventional quantum characteristics of noncommutativity, superposition, and complementarity. And, it entails the invariance of h per the measurement outcomes in inertial reference frames of different complete sets of mutually complementary measurements, which can obviously be justified by the relativity principle.

So, objections to the reconstruction program based on its "highly abstract mathematical assumptions without an immediate physical meaning" and the unjustified empirical fact at its foundation were not unreasonable. However, as shown in this chapter we see that the reconstruction program has produced the quantum-mechanical equivalent of the Lorentz transformations from the light postulate and this allows quantum mechanics to be understood in principle fashion via NPRF + h just like special relativity is understood in principle fashion via NPRF + c [37].

In conclusion, the answer to Wheeler's question [41], "The necessity of the quantum in the construction of existence: out of what deeper requirement does it arise?" according to quantum information theory is NPRF + h, i.e.:

Information Invariance & Continuity as justified by the relativity principle.

This idea is "so simple, so beautiful, so compelling" because it is based on the relativity principle that physicists find persuasive, as we explained in Chapter 1.

Now, let's see how this can be used to derive the joint probabilities for our triplet states in Chapter 2, "to translate the elementary quantum-mechanical reconciliation of cases (a) and (b) into terms meaningful to a general reader struggling with the dilemma raised by the [Mermin] device." That is, in the next chapter we will use NPRF + h to solve the mystery of Bell state entanglement per the Mermin device to answer Mermin's challenge and resolve the EPR and EPR–Bell paradoxes.

References

[1] E. Adlam and C. Rovelli, *Information is Physical: Cross-Perspective Links in Relational Quantum Mechanics*, 2022. Preprint. https://arxiv.org/abs/2203.13342.

[2] P. Ball, *Why Everything You Thought You Knew About Quantum Physics is Different*, 2018. https://www.youtube.com/watch?v=q7v5NtV8v6I.

[3] C. Brukner and A. Zeilinger, *Operationally Invariant Information in Quantum Measurements*, Physical Review Letters, 83 (1999), pp. 3354–3357.

[4] C. Brukner and A. Zeilinger, *Information and fundamental elements of the structure of quantum theory*, in Time, Quantum, Information, L. Castell and O. Ischebeckr, eds., Springer, 2003, pp. 323–354.

[5] C. Brukner and A. Zeilinger, *Information Invariance and Quantum Probabilities*, Foundations of Physics, 39 (2009), pp. 677–689.

[6] J. Bub, *Quantum Mechanics as a Principle Theory*, Studies in History and Philosophy of Science Part B: Studies in History and Philosophy of Modern Physics, 31 (2000), pp. 75–94.

[7] J. Bub, *Bananaworld: Quantum Mechanics for Primates*, Oxford University Press, Oxford, 2016.

[8] J. Bub, *"Two Dogmas" Redux*, in Quantum, Probability, Logic: The Work and Influence of Itamar Pitowski, M. Hemmo and O. Shenker, eds., Springer Nature, London, 2020, pp. 199–215.

[9] J. Bub and I. Pitowski, *Two dogmas about quantum mechanics*, in Many Worlds? Everett, Quantum Theory, and Reality, S. Saunders, J. Barrett, A. Kent, and D. Wallace, eds., Oxford University Press, Oxford, 2010, pp. 431–456.

[10] C. Caves, C. Fuchs, and R. Schack, *Unknown Quantum States: The Quantum de Finetti Representation*, 2001. Preprint. https://arxiv.org/abs/quant-ph/010408.

[11] G. Chiribella, G. M. D'Ariano, and P. Perinotti, *Probabilistic theories with purification*, Physical Review A, 81 (2010), p. 062348.

[12] C. Comte, *General relativity without coordinates*, Nuovo Cimento B, 111 (1996), pp. 937–956.

[13] B. Dakic and C. Brukner, *Quantum Theory and Beyond: Is Entanglement Special?*, in Deep Beauty: Understanding the Quantum World through Mathematical Innovation, H. Halvorson, ed., Cambridge University Press, New York, 2009, pp. 365–392.

[14] O. Darrigol, *Natural Reconstructions of Quantum Mechanics*, in The Oxford Handbook of The History of Quantum Interpretations, O. Freire, ed., Oxford University Press, New York, 2022, pp. 437–472.

[15] P. Evans, *How philosophy turned into physics and reality turned into information*, 2022. https://phys.org/news/2022-10-philosophy-physics-reality.html.

[16] L. Felline, *Mechanisms meet structural explanation*, Synthese, 195 (2018), pp. 99–114.

[17] L. Felline, *Quantum theory is not only about information*, Studies in History and Philosophy of Science Part B: Studies in History and Philosophy of Modern Physics, (2018), pp. 1355–2198.

[18] R. Feynman, *The Character of Physical Law*, MIT Press, Cambridge, MA, 1965.

[19] C. Fuchs, *Quantum Mechanics as Quantum Information (and only a little more)*, 2002. Preprint. https://arxiv.org/abs/quant-ph/0205039.

[20] C. Fuchs and B. Stacey, *Some Negative Remarks on Operational Approaches to Quantum Theory*, in Quantum Theory: Informational Foundations and Foils, G. Chiribella and R. Spekkens, eds., Springer, Dordrecht, 2016, pp. 283–305.

[21] A. Grinbaum, *Elements of information-theoretic derivation of the formalism of quantum theory*, International Journal of Quantum Information, 1 (2003), pp. 289–300.

[22] A. Grinbaum, *Information-theoretic principle entails orthomodularity of a lattice*, Foundations of Physics Letters, 18 (2005), pp. 573–592.

[23] A. Grinbaum, *Reconstruction of quantum theory*, British Journal for the Philosophy of Science, 58 (2007), pp. 387–408.

[24] L. Hardy, *Quantum Theory from Five Reasonable Axioms*, 2001. Preprint. https://arxiv.org/abs/quant-ph/0101012.

[25] L. Hardy, *Reconstructing Quantum Theory*, in Quantum Theory: Informational Foundations and Foils, G. Chiribella and R. Spekkens, eds., Springer, Dordrecht, 2016, pp. 223–248.

[26] P. Höhn, *Toolbox for reconstructing quantum theory from rules on information acquisition*, 2018. Preprint. https://arxiv.org/abs/1412.8323.

[27] P. Höhn, *Complementarity Identities from an Informational Reconstruction*, 2023. Conference: The Quantum Reconstruction Program and Beyond. https://www.youtube.com/watch?v=60ZQ9Fp2cBo.

[28] G. Jaeger, *Information and the Reconstruction of Quantum Physics*, Annalen der Physik, 531 (2018), p. 1800097.

[29] A. Koberinski and M. P. Müller, *Quantum Theory as a Principle Theory: Insights from an Information-Theoretic Reconstruction*, in Physical Perspectives on Computation, Computational Perspectives on Physics, M. E. Cuffaro and S. C. Fletcher, eds., Cambridge University Press, Cambridge, 2018, pp. 257–280.

[30] R. Landauer, *Information is Physical*, Physics Today, 44 (1991), pp. 23–29.

[31] L. Masanes and M. P. Müller, *A derivation of quantum theory from physical requirements*, New Journal of Physics, 13 (2011), p. 063001.

[32] T. Maudlin, *Quantum Non-Locality and Relativity*, Wiley-Blackwell, Oxford, 2011.

[33] N. D. Mermin, *What's Wrong with this Quantum World?*, Physics Today, 57 (2004), pp. 10–11.

[34] M. Müller, 2023. Personal correspondence.

[35] J. Preskill, *Chapter 1: Introduction and Overview*, 2012. http://theory.caltech. edu/preskill/ph229/notes/chap1.pdf.

[36] C. Rovelli, *Relational quantum mechanics*, International Journal of Theoretical Physics, 35 (1996), pp. 1637–1678.

[37] W. M. Stuckey, T. McDevitt, and M. Silberstein, *No Preferred Reference Frame at the Foundation of Quantum Mechanics*, Entropy, 24 (2022), p. 12.

[38] W. M. Stuckey, and M. Silberstein, T. McDevitt, *Completing the Quantum Reconstruction Program via the Relativity Principle*, 2024. https://arxiv.org/abs/2404. 13064.

[39] W. Van Camp, *Principle theories, constructive theories, and explanation in modern physics*, Studies in History and Philosophy of Science Part B: Studies in History and Philosophy of Modern Physics, 42 (2011), pp. 23–31.

[40] S. Weinberg, *The Trouble with Quantum Mechanics*, The New York Review of Books, 19 January (2017). https://www.nybooks.com/articles/2017/01/19/trouble-with-quantum-mechanics/.

[41] J. A. Wheeler, *How Come the Quantum?*, New Techniques and Ideas in Quantum Measurement Theory, 480 (1986), pp. 304–316.

[42] J. A. Wheeler, *Information, physics, quantum: The search for links*, in Proceedings of the 3rd International Symposium on Foundations of Quantum Mechanics in the Light of New Technology, H. Ezawa, S. I. Kobayashi, and Y. Murayama, eds., Physical Society of Japan, Tokyo, 1990, pp. 309–336.

[43] A. Zeilinger, *A Foundational Principle for Quantum Mechanics*, Foundations of Physics, 29 (1999), pp. 631–643.

7

Mystery Solved: Oh, the Irony

[Einstein] behaves now with Bohr exactly as the supporters of absolute
simultaneity behaved with him.

Paul Ehrenfest (3 November 1927).

In the last chapter, we saw that Bohr believed physics in general and quantum mechanics specifically is not about the way Nature is, but what we can say about Nature. Viewing quantum mechanics as a theory about information led to Rovelli and Bub's call for a principle approach to quantum mechanics like that of special relativity, Zeilinger's Foundational Principle for quantum mechanics, and Hardy's discovery that continuity distinguishes quantum probability theory from classical probability theory. All of that culminated in Brukner and Zeilinger's fundamental information-theoretic principle of Information Invariance & Continuity.

Using spin-$\frac{1}{2}$ particles to physically instantiate Information Invariance & Continuity, we saw that it entails the Planck postulate, i.e., everyone measures the same value for h, regardless of their relative spatial orientations. Obviously, the relativity principle can be used to justify the Planck postulate giving us NPRF + h that leads to 'average-only' projection for spin-$\frac{1}{2}$ particles in accord with Dakić and Brukner's closeness requirement, Darrigol's correspondence argument, and Comte's homogeneity of statistical ensembles. This then serves to model the qubit most generally, and quantum information theorists have shown that the entire kinematic structure (Hilbert space) of finite-dimensional quantum mechanics is built upon the qubit.

As Darrigol states for the general reader, quantum information theorists have shown that the kinematic structure of quantum mechanics [5]:

emerges from a natural extension of the lattice of Yes–No empirical questions when the questions do not commute.

Or equivalently, the kinematic structure of quantum mechanics [5]:

results from the harmonious blending of the discontinuity of measurement results with the continuity of the possibilities of measurement.

Einstein's Entanglement. W. M. Stuckey, Michael Silberstein, and Timothy McDevitt, Oxford University Press.
© Oxford University Press (2024). DOI: 10.1093/9780198919698.003.0008

So, we now know that NPRF + h resides at the foundation of quantum mechanics just as NPRF + c resides at the foundation of special relativity.

In this chapter, we will use NPRF + h and 'average-only' projection to solve the mystery of entanglement per the Mermin device in Chapter 2. Specifically, we will derive the joint probabilities of Chapter 2 for the Bell spin triplet state from NPRF + h and the closeness requirement. That will lead us to what is possibly the greatest irony in physics as foreshadowed by Paul Ehrenfest's 1927 quote at the start of the chapter [9, p. 416].

7.1 'Average-Only' Conservation and the Bell States

To remind you, there are processes whereby spin angular momentum is conserved in the emission of a pair of particles. When this happens, the pair of particle spins can be anti-aligned (to sum to zero) or aligned (to sum to $\pm\hbar$). There are four maximally entangled states of this type called *Bell spin states*. The Bell spin singlet state represents an entangled pair of particles with anti-aligned spins in any direction of space. It is written

$$|\psi_-\rangle = \frac{|z+\rangle \otimes |z-\rangle - |z-\rangle \otimes |z+\rangle}{\sqrt{2}}.$$

The symbol $\otimes$ is used to denote the fact that we are talking about two particles, so $|z+\rangle \otimes |z-\rangle$ means particle 1 (Alice's particle, say) is in the state $|\psi\rangle = |z+\rangle$ and particle 2 (Bob's particle, say) is in the state $|\psi\rangle = |z-\rangle$.

This particular method of mathematically combining the states of particles captures exactly what Zeilinger meant when he wrote [18], "For composite systems entanglement results if all possible information is exhausted in specifying joint properties of the constituents." For example, it follows from Axiom 2 of Dakić and Brukner's reconstruction [4]:

(2) (Locality) The state of a composite system is completely determined by measurements on its subsystems.

Similarly for Masanes and Müller's second "physical requirement" in their reconstruction [11]:

2. The state of a composite system is characterized by the statistics of measurements on the individual components.

It is this composition rule that allows one to build the entire Hilbert space starting with the two-dimensional qubit Hilbert space (Figure 6.4). The wavefunction is then constrained to evolve in this Hilbert space according to Schrödinger's equation, i.e., the kinematics constrains the dynamics (Table 5.1). Grinbaum summarized Hardy's approach this way [8]:

> Quantum theory is *about* probabilities, a particular composition rule, and a principle of continuity.

Again, you do not have to worry about the math at all, it suffices to understand conceptually that the singlet state means when Alice and Bob both measure along the same direction $\hat{a} = \hat{b}$ they will always get opposite results.

Likewise, the three Bell spin triplet states,

$$|\psi_+\rangle = \frac{|z+\rangle \otimes |z-\rangle + |z-\rangle \otimes |z+\rangle}{\sqrt{2}},$$

$$|\phi_-\rangle = \frac{|z+\rangle \otimes |z+\rangle - |z-\rangle \otimes |z-\rangle}{\sqrt{2}},$$

$$|\phi_+\rangle = \frac{|z+\rangle \otimes |z+\rangle + |z-\rangle \otimes |z-\rangle}{\sqrt{2}},$$

mean that when Alice and Bob both measure along the same direction in the relevant plane of symmetry they will always get the same results. If you are following the math, you will realize that when Alice and Bob both measure $|\psi_+\rangle$ along $\hat{z}$ they will always get opposite results. That does not contradict what we said about triplet states because $\hat{z}$ is not in the symmetry plane of $|\psi_+\rangle$, which is the *xy* plane.

The triplet state in its symmetry plane is the situation represented by the Mermin device. [Although, Mermin prefers having R (G) mean +1 (−1) for Alice and −1 (+1) for Bob, so the Mermin device can represent the singlet state. The explanation that follows is equally valid either way.] The three choices of detector settings on the Mermin device correspond to three different SG magnet orientations in the plane orthogonal to the particle beams from the source (Figure 2.6), where Alice's SG magnet orientation is given by $\hat{a}$ and Bob's is given by $\hat{b}$ (Figure 2.5).

Recall the joint probabilities we need to derive in order to answer Mermin's challenge and solve the mystery of Bell state entanglement. First, the probability of Alice and Bob obtaining the same results as a function of the angle θ between $\hat{a}$ and $\hat{b}$ for the triplet state is

$$P(++) = P(--) = \frac{1}{2}\cos^2(\theta/2).$$

Second, the probability of Alice and Bob obtaining different results as a function of the angle θ between $\hat{a}$ and $\hat{b}$ for the triplet state is

$$P(+-) = P(-+) = \frac{1}{2}\sin^2(\theta/2).$$

These provide "the elementary quantum-mechanical reconciliation of cases (a) and (b)" for the Mermin device.

That is, Fact 1 for case (a) says Alice and Bob always get the same outcomes (both get G or both get R) when they happen to select the same settings (both happen to select setting 1, or both happen to select setting 2, or both happen to select setting 3). Fact 2 for case (b) says Alice and Bob get the same outcomes in $\frac{1}{4}$ of the trials when they happen to select different settings. Same settings means $\theta = 0$ in our triplet state joint probabilities and different settings means $\theta = 120°$ in our triplet state joint probabilities (Figure 2.6). That is how our triplet state joint probabilities provide "the elementary quantum-mechanical reconciliation of cases (a) and (b)" for the Mermin device. Given Information Invariance & Continuity as justified by the relativity principle (NPRF + h) and the closeness requirement, these joint probabilities are easy to derive.

Suppose Alice obtains +1 at $\hat{a}$ and Bob measures at $\hat{b} \neq \hat{a}$ ($\theta \neq 0$). We have the exact same situation here between $\hat{a}$ and $\hat{b}$ for two particles that we had in Chapter 6 between $\hat{z}$ and $\hat{b}$ for one particle. Using the same reasoning here that we used in Chapter 6, Alice says Bob's measurement outcome should be $\cos(\theta)$, since obviously he would also have gotten +1 for his particle if he had measured at $\hat{b} = \hat{a}$, as required to conserve spin angular momentum (Figure 7.1). The problem is, again, that would mean Alice alone measures h while Bob measures some fraction of h, which means Alice occupies a preferred reference frame. Since Bob must also always measure h per NPRF, Bob's ± 1 outcomes can only *average* to $\cos(\theta)$ at best (Figure 7.2). That means, from Alice's perspective, Bob's measurement outcomes only satisfy conservation of spin angular momentum *on average* when Bob is measuring the spin of his particle in a different inertial reference frame.

We can write this 'average-only' conservation for Alice's +1 outcomes as

$$2P(++)(+1) + 2P(+-)(-1) = \cos(\theta).$$

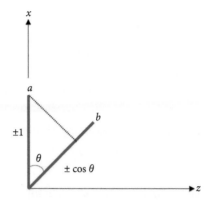

Fig. 7.1 Per Alice, Bob should be measuring $\pm \cos(\theta)$ when she measures ± 1, respectively.

Fig. 7.2 Average view for the triplet state. Reading from left to right, as Bob rotates his SG magnets (rotating blue arrow) relative to Alice's SG magnets (blue arrow always vertically oriented) for her +1 outcome (the black dot at the tip of her arrow), the average value of his outcome (the black dot along his arrow) varies from +1 (totally up, arrow tip) to 0 to –1 (totally down, arrow bottom). This obtains per conservation of spin angular momentum on average in accord with NPRF. Bob can say exactly the same about Alice's outcomes as she rotates her SG magnets relative to his SG magnets for his +1 outcome.

Likewise, for Alice's –1 outcomes 'average-only' conservation is written

$$2P(-+)(+1) + 2P(--)(-1) = -\cos(\theta).$$

This 'average-only' conservation plus normalization per NPRF,

$$P(++) + P(+-) = \frac{1}{2}, \qquad P(-+) + P(--) = \frac{1}{2},$$

means $P(++) = P(--) = \frac{1}{2}\cos^2(\theta/2)$ and $P(+-) = P(-+) = \frac{1}{2}\sin^2(\theta/2)$, as required by "the elementary quantum-mechanical reconciliation of cases (a) and (b)" for the Mermin device.

As Darrigol points out [5], the empirical fact called spin does not provide a compelling starting point to derive the qubit probabilities leading to these triplet state joint probabilities. What he is lacking is the compelling fundamental principle that justifies the empirical fact, i.e., the relativity principle. Then and only then do we have the principle explanation shown in Chapter 1:

compelling fundamental principle	$\rightarrow$	justifies empirically discovered fact	$\rightarrow$	dictating the mystery

We should point out that the trial-by-trial outcomes for this 'average-only' conservation can deviate substantially from the target value required for explicit conservation per Alice's reference frame. For example, we might have Bob's +1 and –1 outcomes averaging to zero as required for the conservation of spin angular momentum per Alice's reference frame (Figure 7.3). Consequently, Alice says Bob's measurement outcomes are violating the conservation of spin angular momentum as egregiously as possible on a trial-by-trial basis. In classical physics, our conservation laws hold on average because they hold explicitly for each and every trial of the experiment (within experimental limits). But here, that would require that Bob measure a fraction of h while Alice measures h proper, meaning Alice would occupy a preferred reference frame.

So, 'average-only' conservation distinguishes classical mechanics and quantum mechanics just as length contraction distinguishes Newtonian mechanics and special relativity. Consequently, we see that 'average-only' conservation does not *solve* the mystery of entanglement, it *is* the mystery, i.e., it is what

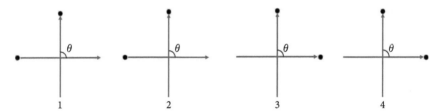

Fig. 7.3 An ensemble of four SG measurement trials of a spin triplet state showing Bob's (Alice's) outcomes (black dots on horizontal arrows) corresponding to Alice's (Bob's) +1 outcome (black dots on vertical arrows) when $\theta = 90°$. Spin angular momentum is not conserved in any given trial, because there are two different measurements being made, i.e., the outcomes are in two different reference frames, but it is conserved *on average* for all four trials (two down outcomes and two up outcomes average to $\cos{(90°)} = 0$).

needs to be explained [17]. The relativity principle and Information Invariance & Continuity provide us with that explanation, i.e., NPRF + h.

7.2 Relativity of Data Partition

But, wait a minute. Why is it that Bob has to average *his* results to accommodate Alice? The answer is, of course, that he does not. The data are perfectly symmetrical with respect to Alice and Bob, so Bob can partition the data according to his +1 and –1 results and argue equally that it is Alice who must average *her* data in accord with the conservation of spin angular momentum (Figures 7.2 and 7.4). This is exactly analogous to the relativity of simultaneity for special relativity in Chapter 4. There, we saw that Alice would foliate (partition) events in spacetime according to her surfaces of simultaneity and say that Bob's meter sticks are short and his clocks run slow, while Bob would foliate (partition) events in spacetime according to his surfaces of simultaneity and say that it is Alice's meter sticks that are short and her clocks that run slow (Table 7.1).

All of these arguments can be applied equally to the singlet state, as Mermin preferred. There, the probabilities to be derived are reversed: $P(+-) = P(-+) = \frac{1}{2}\cos^2(\theta/2)$ and $P(++) = P(--) = \frac{1}{2}\sin^2(\theta/2)$. Since Bob obtains –1 when Alice obtains +1 for $\hat{b} = \hat{a}$ and vice versa for the singlet state, 'average-only' conservation for Alice's +1 outcomes becomes

$$2P(++)(+1) + 2P(+-)(-1) = -\cos(\theta),$$

while for Alice's –1 outcomes 'average-only' conservation becomes

$$2P(-+)(+1) + 2P(--)(-1) = \cos(\theta).$$

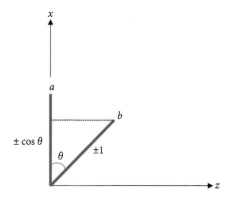

Fig. 7.4 Per Bob, Alice should be measuring $\pm\cos(\theta)$ when he measures ± 1, respectively.

Table 7.1 Principle comparison of special relativity and quantum mechanics. Because Alice and Bob both measure the same speed of light c, regardless of their motion relative to the source per NPRF, Alice (Bob) may claim that Bob's (Alice's) length and time measurements are erroneous and need to be corrected (length contraction and time dilation). Likewise, because Alice and Bob both measure the same values for spin angular momentum ±1 $(\hbar/2)$ regardless of their SG magnet orientation relative to the source per NPRF, Alice (Bob) may claim that Bob's (Alice's) individual ±1 values are erroneous and need to be corrected (averaged, Figures 7.2 and 7.3).

	Empirical fact	Consequence
Special relativity	Alice and Bob both measure c, regardless of their motion relative to the source	Alice (Bob) says Bob (Alice) must correct his (her) time and length measurements
Quantum mechanics	Alice and Bob both measure ±1 $(\hbar/2)$, regardless of their SG orientation relative to the source	Alice (Bob) says Bob (Alice) must average his (her) ±1 results

Fig. 7.5 Average view for the singlet state. Compare with Figure 7.2.

Normalization per NPRF is the same, so normalization and these properly revised 'average-only' conservation equations do in fact give the singlet state joint probabilities. Consequently, Figure 7.2 now looks like Figure 7.5.

So, Information Invariance & Continuity as justified by the relativity principle satisfies Rovelli's desideratum exactly as he articulated it in 1996 with his introduction to relational quantum mechanics [14]. There, he not only likened his two axioms for quantum mechanics (Section 6.6) to the two postulates of special relativity, but he likened the relativity of simultaneity to the idea that quantum states only take values relative to a particular experimental context. NPRF + h turns his analogy into a reality.

7.3 Compatibility of Quantum Mechanics and Special Relativity

But, wait a minute. This all looks like quantum mechanics and special relativity are highly compatible (Figure 7.6 and Table 7.1). Does this not violate conventional wisdom? In 2018, Marco Mamone-Capria even published a paper titled [10] "On the Incompatibility of Special Relativity and Quantum Mechanics." There he writes:

> An important feature of the theoretical landscape in physics during the last ninety years is that the two main foundational theories, relativity and quantum mechanics, at least in their standard interpretations, contradict each other. Basically, this is not even a very subtle remark, since the Schrödinger equation, one of the cornerstones of quantum mechanics, is patently not [Lorentz] invariant. It would have been quite surprising if researchers had never come across experimental settings where quantum mechanics is at variance with special relativity, and completely unrealistic that they had not become aware rather soon of the inconsistency issue.

Again in 1986, Bell wrote [2]:

> For me this is the real problem with quantum theory: the apparently essential conflict between any sharp formulation and fundamental relativity. ... It may be that a real synthesis of quantum and relativity theories requires not just technical developments but radical conceptual renewal.

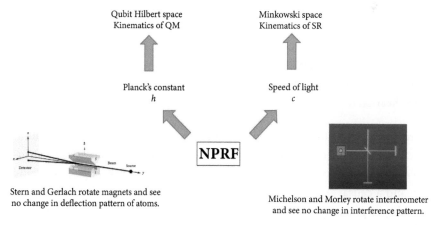

Stern and Gerlach rotate magnets and see no change in deflection pattern of atoms.

Michelson and Morley rotate interferometer and see no change in interference pattern.

Fig. 7.6 NPRF at the foundation of quantum mechanics and special relativity.

And in 1972, Paul Dirac wrote [7, p. 11]:

> The only theory which we can formulate at the present is a non-local one, and of course one is not satisfied with such a theory. I think one ought to say that the problem of reconciling quantum theory and relativity is not solved.

The list goes on. But, as we pointed out in Chapter 3, it simply cannot be the case that entanglement is incompatible with Lorentz invariance because both are experimentally vindicated in their respective realms of applicability, and quantum mechanics is just the low-energy approximation to relativistic (Lorentz-invariant) quantum field theory.

By analogy, Newtonian mechanics differs from special relativity in that it is not Lorentz invariant. Accordingly, Newtonian mechanics predicts Galilean velocity transformation as opposed to the Lorentz velocity transformation of special relativity (as we pointed out in Chapter 4). So, imagine we found experimentally that velocities add according to Galilean rather than Lorentz velocity transformation at all speeds. That would not only mean Newtonian mechanics and special relativity are at odds, it would mean special relativity has been empirically refuted. Most physicists do not believe that entanglement empirically refutes special relativity, so what is prompting the conventional wisdom?

In their 2009 paper "Was Einstein Wrong?: A Quantum Threat to Special Relativity," David Albert and Rivka Galchen write [1]:

> But entanglement also appears to entail the deeply spooky and radically counterintuitive phenomenon called nonlocality—the possibility of physically affecting something without touching it or touching any series of entities reaching from here to there. Nonlocality implies that a fist in Des Moines can break a nose in Dallas without affecting any other physical thing (not a molecule of air, not an electron in a wire, not a twinkle of light) anywhere in the heartland.

Here we see that the reason for the perceived conflict between quantum mechanics and special relativity resides in adherence to Reichenbach's Principle. That is, the reason for the conventional wisdom is that people tacitly demand a causal account of the Bell state correlations. Indeed, this nonlocal causality is precisely what Einstein derided as "spooky actions at a distance." In other words, Einstein was flummoxed by entanglement precisely because he also adhered to Reichenbach's Principle and constructive explanation (see Chapter 5), which brings us to Ehrenfest's 1927 quote at the beginning of this chapter and possibly the greatest irony in physics.

Here is something Bohr himself wrote [12]:

> Indeed from our present standpoint, physics is to be regarded not so much as the study of something a priori given, but as the development of methods for ordering and surveying human experience.

This attitude applied to quantum mechanics led to the principle solution of the greatest mystery in physics just shown, i.e., NPRF + h. When Einstein complained that quantum mechanics is missing "elements of reality," i.e., "something a priori given," he was complaining that quantum mechanics is not a constructive theory while Bohr was arguing that we should not expect physics to provide constructive accounts. Bohr even said that physics provides "the development of methods for ordering and surveying human experience" exactly in accord with Einstein's phenomenological view of physics whereby "the totality of our sense experiences is such that by means of thinking (operations with concepts, and the creation and use of definite functional relations between them, and the coordination of sense experiences to these concepts) it can be put in order" (Chapter 1).

As we said in Chapter 1, the principle account of entanglement presented here avoids violations of locality, statistical independence, intersubjective agreement, and unique experimental outcomes associated with constructive accounts because dynamical laws and mechanistic causal processes are never invoked. And, if principle explanation is actually fundamental to constructive explanation as we argued in Chapter 5, then it is possible that no constructive counterpart to this principle explanation of entanglement is forthcoming. This is exactly what has happened with special relativity's principle explanation of length contraction and thus we have the greatest irony in physics.

That is, Bohr's attitude towards entanglement mirrored exactly Einstein's attitude towards length contraction. There, Einstein abandoned "constructive efforts" to solve that mystery with NPRF + c, giving us relative simultaneity instead of Newton's absolute simultaneity. So, it is supremely ironic that when it came to his own EPR paradox, he refused to abandon constructive efforts, instead choosing allegiance to Reichenbach's Principle over his beloved relativity principle. In his defense, that was years before Bell's "most profound discovery of science."

But, that brings us to yet another irony. As we saw above, Bell subscribed to conventional wisdom that quantum mechanics is at odds with special relativity. Therefore, it is ironic that in 1990, when asked if he thought the mystery of entanglement would ever be solved, Bell responded [3, p. 85]:

> I think the problems and puzzles we are dealing with here will be cleared up, and ... our descendants will look back on us with the same kind of superiority as we now are tempted to feel when we look at people in the late nineteenth century who worried about the aether. And Michelson–Morley ... the puzzles seemed insoluble to them. And came Einstein in nineteen five, and now every schoolboy learns it and feels ... superior to those old guys. Now, it's my feeling that all this action at a distance and no action at a distance business will go the same way. But someone will come up with the answer, with a reasonable way of looking at these things. If we are lucky it will be some big new development like the theory of relativity.

Oh, the irony of it all.

7.4 Qubit for Photon Polarization

In conclusion, Information Invariance & Continuity as justified by the relativity principle gives us NPRF + h and leads to 'average-only' conservation whence the joint probabilities for "the elementary quantum-mechanical reconciliation of cases (a) and (b)." So, NPRF + h translates the quantum probabilities "into terms meaningful to a general reader struggling with the dilemma raised by the [Mermin] device" without violating locality, statistical independence, intersubjective agreement, or unique experimental outcomes, thereby answering Mermin's challenge. In what is perhaps the greatest irony in physics, the EPR and EPR–Bell paradoxes aka "Einstein's Quantum Riddle" [13] are resolved with Einstein's own relativity principle.

Of course, there is nothing unique about SG spin-$\frac{1}{2}$ measurements except that they can be considered direct measurements of h, and that is because we are dealing with h as an action per angular momentum multiplied by angular distance, and angular distance is dimensionless. The more general combination of NPRF and Information Invariance & Continuity described in Chapter 6 applies to *any* qubit. So, for example, it should also apply to spin-1 photons passing or not passing through a polarizer. It does, and the reasoning mirrors that for spin-$\frac{1}{2}$.

As we pointed out in Chapter 2, Planck introduced h in his explanation of blackbody radiation. Accordingly, we now understand that electromagnetic radiation with frequency f is comprised of indivisible quanta (photons) of energy hf. One difference between the classical view of a continuous electromagnetic field and the quantum reality of photons is manifested in polarization measurements.

The axis along which the electric field $\vec{E}$ oscillates linearly in the electromagnetic wave is the wave's linear polarization, and a polarizer has a specific axis along which the electric field is allowed to oscillate without interacting with the polarizer. This allows $\vec{E}$ along that axis to pass through the polarizer unimpeded (Figure 7.7). So, when the polarizing axis of the polarizer is aligned with the polarization of the wave, all of the incident electric field is transmitted. And, when the electric field makes an angle ϕ with the polarizing axis, only the component $E \cos(\phi)$ of $\vec{E}$ along the polarizing axis is transmitted. Since the energy of the wave is proportional to E^2, the amount of energy that the polarizer transmits goes as $\cos^2(\phi)$ (Figure 7.7).

According to classical electromagnetism, the energy of polarized electromagnetic radiation transmitted by a polarizer can be made as small as you like by continuously varying ϕ. However, given that the radiation is actually composed of indivisible photons, there is a nonzero lower limit to the energy passed by a polarizer, i.e., each quantum of energy hf either passes or does not.

For example, if vertically polarized light is passed through a polarizer angled at 45° with respect to the vertical, classical electromagnetism says half the light

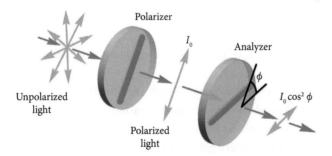

Fig. 7.7 Unpolarized light incident on a polarizer has its intensity (power per unit area) reduced to I_0 and becomes polarized along the polarizer's polarization (transmission) axis. A polarizer placed immediately thereafter with its polarization axis making an angle ϕ with the polarization of the wave further reduces the intensity to $I_0 \cos^2(\phi)$ and polarizes the transmitted light along its polarization axis.

will pass. So, when we send vertically polarized photons to the polarizer, half will pass and half will not. If 'half photons' passed through the polarizer in accord with classical physics, then their energy on the other side would be given by $E = \frac{h}{2}f$ instead of $E = hf$, meaning the value of Planck's constant was effectively cut in half [19].

That means the classical 'expectation' of fractional amounts of quanta can only obtain on average per Information Invariance & Continuity. That is, we would have 'average-only' transmission of energy for spin-1 photons instead of 'average-only' projection of spin angular momentum for spin-$\frac{1}{2}$ particles, both of which give 'average-only' conservation between different inertial reference frames related by rotations. Shinichi Saito writes [15]:

> The theory of relativity established the principle that the speed of light, thus, the momentum of a photon in a vacuum is invariant under the Lorentz trans-formation. *The principle behind our SU(2) theory of a photon is the rotational symmetry of the angular momentum of a photon in a vacuum.* In other words, there is no particular preferential polarisation state for a photon to be realised in a vacuum, no matter which direction the photon is propagating.

So, most generally, NPRF and the information-theoretic principle of Information Invariance & Continuity lead to 'average-only' —— (fill in the blank) giving 'average-only' conservation of the measured quantity for its Bell states. For example, the Bell state in "Entangled photons, nonlocality and Bell inequalities in the undergraduate laboratory" is [6]:

$$|\psi_{\text{EPR}}\rangle = \frac{|V\rangle \otimes |V\rangle + |H\rangle \otimes |H\rangle}{\sqrt{2}},$$

where V stands for vertical (pass) along some axis and H stands for horizontal (no pass) along that same axis.

Circular polarization is associated with spin for the photon (spin angular momentum of $\pm\hbar$) and the complementary variables are S_x, S_y, and S_z with corresponding reference frames à la spin-$\frac{1}{2}$, i.e., the Bloch sphere (Figure 6.3) for spin-$\frac{1}{2}$ becomes the Poincaré sphere for photon polarization [15]. The linear polarization states $|V\rangle$ and $|H\rangle$ that we will deal with are superpositions of the circular polarization states, so they also reside on the Poincaré sphere. We will introduce the complementary measurements for linear photon polarization in Chapter 9.

Notice that the photon qubit differs from the spin-$\frac{1}{2}$ qubit in two ways. First, ϕ in real space equals ϕ in Hilbert space for the photon qubit [6, 16] (Figure 7.8) while θ in real space gives $\theta/2$ in Hilbert space for the spin-$\frac{1}{2}$ qubit (Figure 6.4). Second, the photon qubit is giving a binary distribution of quantum events in time (rate of "pass" clicks relative to "no pass" nonclicks in the photon detector behind the polarizer) while the spin-$\frac{1}{2}$ qubit is giving a binary distribution of quantum events in a direction of space (number of "up" detector clicks relative to "down" detector clicks relative to the SG magnetic field).

Formally, if $|\psi\rangle = |V$ for $z\rangle$ and we rotate the polarizer from being along the z axis to being along the x axis (notice the polarizer does not select a direction along its orientation like the SG magnets), the number of photons passed per second decreases from a maximum to zero (Figure 7.8). That is, $|H$ for $z\rangle$ is $|V$ for $x\rangle$. So, we are dealing with h as an action according to energy multiplied by time in this experiment.

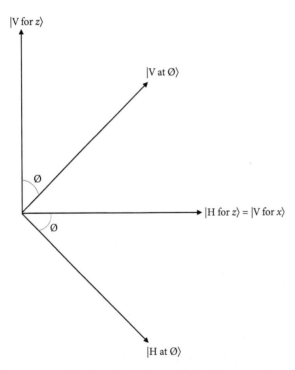

Fig. 7.8 Hilbert space qubits for the photon polarization experiment. The basis for a polarization measurement along the z axis is $|V$ for $z\rangle$ (pass, vertical) and $|H$ for $z\rangle$ (no pass, horizontal). The basis for a polarization measurement at angle ϕ relative to the z axis is $|V$ at $\phi\rangle$ and $|H$ at $\phi\rangle$. The angle ϕ in real space equals the angle between qubits in Hilbert space, so $|H$ for $z\rangle = |V$ for $x\rangle$.

7.5 Summary

In summary, a principle solution to the greatest mystery in physics is:

relativity principle → Planck postulate → Bell state correlations

as entailed by NPRF and the information-theoretic principle of Information Invariance & Continuity. More specifically, the relativity principle (NPRF) is a compelling fundamental principle per Einstein's phenomenological view of physics (Chapter 1). In that context, it can be stated broadly for the general reader as:

No empirical data provide a privileged perspective on the real external world.

This compelling fundamental principle demands that everyone measure the same value for any constant of Nature, which in this case is Planck's constant h. That is, the relativity principle justifies the Planck postulate whence the empirically discovered quantum property of spin angular momentum that dictates 'average-only' projection giving the qubit probabilities for spin-$\frac{1}{2}$, as well as 'average-only' transmission giving the qubit probabilities for spin-1 photons. In Subchapter 9C, we will see that NPRF + h dictates wave–particle duality giving the qubit probabilities for the double-slit experiment. This is all in accord with the standard quantum-mechanical concepts of noncommutativity, superposition, and complementarity, i.e., quantum mechanics does not need to be 'fixed' according to our principle account.

When two particles are spin entangled, 'average-only' projection (or transmission for spin-1 photons) becomes 'average-only' conservation giving the otherwise mysterious joint probabilities for the Mermin device. So, 'average-only' conservation per NPRF and Information Invariance & Continuity explains Fact 1 for case (a) and Fact 2 for case (b) for the Mermin device. This is how the information-theoretic reconstructions of quantum mechanics (completed via principle explanation) allow us to solve the mystery of entanglement so as to resolve the EPR and EPR–Bell paradoxes and answer Mermin's challenge with the compelling fundamental principle NPRF.

And, since quantum information theorists have shown that quantum-mechanical Hilbert space is built on the qubit, we know that NPRF + h resides at the foundation of quantum mechanics just as NPRF + c resides at the foundation of special relativity. Very few physicists complain that "nobody understands special relativity," so it is probably safe to say that thanks to the efforts of quantum information theorists beginning with Rovelli (1996) and proceeding through Zeilinger (1999), Brukner (1999), Fuchs (2000), Hardy (2001), Bub (2003), Clifton (2003), Halvorson (2003), Grinbaum (2005), Dakić (2009), Masanes (2011), Müller (2011), Höhn (2014), and many others, we now understand quantum mechanics.

In the next chapter, we will bring this to bear on another big question in quantum information theory: does Nature harbor even stronger correlations than those of quantum entanglement?

References

[1] D. Albert and R. Galchen, *Was Einstein Wrong? A Quantum Threat to Special Relativity*, Scientific American, 1 March (2009). https://www.scientificamerican.com/article/was-einstein-wrong-about-relativity/.
[2] J. Bell, *Introductory Remarks*, Physics Reports, 137 (1986), pp. 7–9.
[3] J. Bell, *Indeterminism and Nonlocality*, in Mathematical Undecidability, Quantum Nonlocality and the Question of the Existence of God, A. Driessen and A. Suarez, eds., Springer, Dordrecht, 1997, pp. 78–89.
[4] B. Dakic and C. Brukner, *Quantum Theory and Beyond: Is Entanglement Special?*, in Deep Beauty: Understanding the Quantum World through Mathematical Innovation, H. Halvorson, ed., Cambridge University Press, New York, 2009, pp. 365–392.
[5] O. Darrigol, *Natural Reconstructions of Quantum Mechanics*, in The Oxford Handbook of The History of Quantum Interpretations, O. Freire, ed., Oxford University Press, New York, 2022, pp. 437–472.
[6] D. Dehlinger and M. Mitchell, *Entangled photons, nonlocality, and Bell inequalities in the undergraduate laboratory*, American Journal of Physics, 70 (2002), pp. 903–910.
[7] P. A. M. Dirac, *Development of the Physicist's Conception of Nature*, in The Physicist's Conception of Nature, J. Mehra, ed., Reidel, Dordrecht, 1973, pp. 1–14.
[8] A. Grinbaum, *Quantum Information and the Quest for Reconstruction of Quantum Theory*, in The Oxford Handbook of The History of Quantum Interpretations, O. Freire, ed., Oxford University Press, New York, 2022, pp. 417–436.
[9] J. Kalckar, *Neils Bohr: Collected Works, Volume 6 – Foundations of Quantum Physics I (1926–1932)*, North-Holland, Amsterdam, 1985.
[10] M. Mamone-Capria, *On the Incompatibility of Special Relativity and Quantum Mechanics*, Journal for Foundations and Applications of Physics, 8 (2018), pp. 163–189.
[11] L. Masanes and M. P. Müller, *A derivation of quantum theory from physical requirements*, New Journal of Physics, 13 (2011), p. 063001.
[12] N. D. Mermin, *What's Wrong with this Quantum World?*, Physics Today, 57 (2004), pp. 10–11.
[13] NOVA Season 46 Episode 2, *Einstein's Quantum Riddle*, 2019. https://www.pbs.org/video/einsteins-quantum-riddle-ykvwhm/.
[14] C. Rovelli, *Relational quantum mechanics*, International Journal of Theoretical Physics, 35 (1996), pp. 1637–1678.
[15] S. Saito, *Spin of Photons: Nature of Polarisation*, 2023. Preprint. https://arxiv.org/abs/2303.17112v1.

[16] W. M. Stuckey, M. Silberstein, T. McDevitt, and I. Kohler, *Why the Tsirelson Bound? Bub's Question and Fuchs' Desideratum*, Entropy, 21 (2019), p. 692.

[17] W. M. Stuckey, M. Silberstein, T. McDevitt, and T. D. Le, *Answering Mermin's challenge with conservation per no preferred reference frame*, Scientific Reports, 10 (2020), p. 15771.

[18] A. Zeilinger, *A Foundational Principle for Quantum Mechanics*, Foundations of Physics, 29 (1999), pp. 631–643.

[19] B. Zwiebach. *Photons and the loss of determinism* (2017). https://www.youtube.com/watch?v=8OsUQ1yXCcI.

8

Superquantum Probabilities: Why No PR-Box?

The vast majority of attempts to find physical principles behind quantum theory either fail to single out the theory uniquely or are based on highly abstract mathematical assumptions without an immediate physical meaning.

Borivoje Dakić and Časlav Brukner (2009)

In Chapters 6 and 7, we saw how the invariant spin measurement of Planck's constant h per Information Invariance & Continuity as justified by the relativity principle solved the mystery of Bell state entanglement. That is, NPRF + h necessitates the 'average-only' conservation responsible for the EPR–Bell paradox. This ruled out the classical constructive model of spin which led to only one reference frame measuring h while every other reference frame measured some fraction of h, i.e., the classical account of spin led to a preferred reference frame. Accordingly, the relativity principle dictates quantum instead of classical behavior for the conservation of spin angular momentum.

In terms of quantum information theory, NPRF + h is responsible for the fact that there is a limit to how much simultaneous information is available for a quantum system. In this chapter, we will use that result to respond to Dakić and Brukner above [5] and to "single out [quantum mechanics] uniquely" from the so-called superquantum joint probabilities of Sandu Popescu and Daniel Rohrlich, i.e., the *PR-box* [7].

8.1 The No-Signaling PR-Box

In 1994, Popescu and Rohrlich introduced their joint probabilities (PR-box) that do not permit faster-than-light signaling (they are no-signaling joint probabilities), but violate Bell's inequality beyond quantum mechanics. As

Einstein's Entanglement. W. M. Stuckey, Michael Silberstein, and Timothy McDevitt, Oxford University Press.
© Oxford University Press (2024). DOI: 10.1093/9780198919698.003.0009

we will see, the PR-box permits amazing information transfer even if it is no-signaling, so quantum information theorists want to know if the PR joint probabilities are realized in Nature. Here is the PR-box:

$$P(+1, +1 \mid a, b) = P(-1, -1 \mid a, b) = \frac{1}{2},$$

$$P(+1, +1 \mid a, b') = P(-1, -1 \mid a, b') = \frac{1}{2},$$

$$P(+1, +1 \mid a', b) = P(-1, -1 \mid a', b) = \frac{1}{2},$$

$$P(+1, -1 \mid a', b') = P(-1, +1 \mid a', b') = \frac{1}{2}.$$

These represent two possible outcomes [+1, −1] for Alice and Bob when a', a are Alice's measurement settings and b', b are Bob's measurement settings. First, let's show that the PR-box satisfies no signaling.

No signaling means that, for a given experimental setting, Alice's (Bob's) outcome is independent of Bob's (Alice's) setting. For example, suppose that Alice chooses setting a. Then no signaling implies that her observed outcome (+1 or −1) is independent of whether Bob chooses setting b or b'. Specifically, the probability that Alice observes +1 given that she chooses setting a and Bob chooses setting b is $P(+1, +1 \mid a, b) + P(+1, -1 \mid a, b) = \frac{1}{2} + 0 = \frac{1}{2}$, and this must equal the probability that she observes +1 if Bob chooses b' instead, $P(+1, +1 \mid a, b') + P(+1, -1 \mid a, b') = \frac{1}{2} + 0 = \frac{1}{2}$. Clearly, the PR-box satisfies no signaling.

The joint probabilities for our Bell states are also no signaling. To see that, start with $P(+1, +1 \mid a, b) + P(+1, -1 \mid a, b) = \frac{1}{2} \cos^2 (\theta/2) + \frac{1}{2} \sin^2 (\theta/2) = \frac{1}{2}$ for the probability that Alice measures +1 for any value of θ. Since θ determines Bob's measurement setting given Alice's, we see that Alice's outcome is indeed independent of Bob's setting. In other words, Alice measures +1 half the time in any measurement direction (so she measures −1 half the time), regardless of what direction Bob is measuring.

If Alice saw different outcomes based on Bob's measurements, Bob could send her signals faster than light, since the outcomes can be spacelike separated (Chapter 4). Given the ambiguous temporal ordering of spacelike separated events, this leads to ambiguous temporal ordering of cause and effect, as described in Chapter 4. That is why the no-signaling property is very important for information theory. However, as we will see, the no-signaling

PR-box still manages to produce information transfer with 'unreasonable' effectiveness.

The reason for this unreasonableness is that the no-signaling joint probabilities of the PR-box produce stronger measurement correlations than those of quantum mechanics, i.e., they violate Bell's inequality to a greater degree than quantum mechanics. To see that, we need another form of Bell's inequality called the CHSH inequality alluded to in Chapter 0.

8.2 The CHSH Bell Inequality

In 1969, John Clauser, Michael Horne, Abner Shimony, and Richard Holt constructed a more experiment-friendly version of Bell's inequality that now bears their names, i.e., the Clauser–Horne–Shimony–Holt (CHSH) inequality [4]. Here is Mermin's 2005 derivation of that inequality [6].

Suppose Alice and Bob restrict themselves to making two measurements each, i.e., Alice makes measurements a and d while Bob makes measurements b and c. We will again assume the particles have instruction sets that dictate their outcomes prior to any possible measurement in accord with local realism (recall this from Chapter 2). Let the instruction set for Alice's particle in trial j of the measurement sequence be written $A_a(j)$ for her measurement a and $A_d(j)$ for her measurement d. Similarly, the instruction set for Bob's particle in trial j of the measurement sequence is written $B_b(j)$ for his measurement b and $B_c(j)$ for his measurement c. Now define the quantity

$$\gamma_j := A_a(j)B_b(j) + A_a(j)B_c(j) + A_d(j)B_b(j) - A_d(j)B_c(j).$$

Since $A_x(j) = \pm 1$ and $B_y(j) = \pm 1$ (x = a or d, y = b or c), γ_j is either $+2$ or -2 for any j. So, the average of γ_j over many trials (called γ) is confined to the range $[-2, +2]$ for our classical constructive account, i.e., for instruction sets. This is the CHSH inequality.

Since Alice and Bob can each only make a single measurement for each trial j, what we obtain experimentally is an average for each of the four terms in γ_j:

$$\gamma_{\text{exp}} = \langle A_a B_b \rangle + \langle A_a B_c \rangle + \langle A_d B_b \rangle - \langle A_d B_c \rangle.$$

If Alice and Bob are making their spin measurements on a Bell spin triplet state, then quantum mechanics predicts

$$\langle A_x B_y \rangle = (+1)(+1)\frac{1}{2}\cos^2\left(\frac{\theta_{xy}}{2}\right) + (-1)(-1)\frac{1}{2}\cos^2\left(\frac{\theta_{xy}}{2}\right) +$$

$$(+1)(-1)\frac{1}{2}\sin^2\left(\frac{\theta_{xy}}{2}\right) + (-1)(+1)\frac{1}{2}\sin^2\left(\frac{\theta_{xy}}{2}\right) = \cos\left(\theta_{xy}\right).$$

For the Bell spin singlet state, $\cos \rightarrow \sin$ and $\sin \rightarrow \cos$ giving $\langle A_x B_y \rangle = -\cos\left(\theta_{xy}\right)$. So, for measurements on $|\phi_+\rangle$ in its symmetry (xz) plane (Figure 2.5) we have

$$\gamma_{\exp} = \cos(\theta_{ab}) + \cos(\theta_{ac}) + \cos(\theta_{db}) - \cos(\theta_{dc})$$

Using the measurements shown in Figure 8.1 where $\theta_{ab} = \theta_{ac} = \theta_{db} = 45°$ and $\theta_{dc} = 135°$, we see that quantum mechanics predicts $\gamma_{\exp} = 2\sqrt{2}$. Using these same measurements on the singlet state gives $\gamma_{\exp} = -2\sqrt{2}$. These limits are called the *Tsirelson bound* because in 1980 Boris Tsirelson (also spelled Cirel'son) showed that they are the largest possible violations of the CHSH inequality for quantum mechanics [3].

Now let's see what the PR-box gives for $\gamma_{\exp}$. Letting $c = b'$ and $d = a'$, the PR-box joint probabilities give

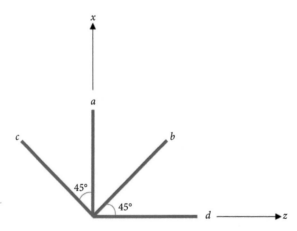

Fig. 8.1 Alice and Bob's CHSH inequality-violating measurements of a Bell spin state.

$$\langle A_a B_b \rangle = (+1)(+1)P(+1,+1 \mid a,b) + (-1)(-1)P(-1,-1 \mid a,b)+$$
$$(+1)(-1)P(+1,-1 \mid a,b) + (-1)(+1)P(-1,+1 \mid a,b)$$
$$= \frac{1}{2} + \frac{1}{2} - 0 - 0 = 1,$$

$$\langle A_a B_c \rangle = (+1)(+1)P(+1,+1 \mid a,b') + (-1)(-1)P(-1,-1 \mid a,b')+$$
$$(+1)(-1)P(+1,-1 \mid a,b') + (-1)(+1)P(-1,+1 \mid a,b')$$
$$= \frac{1}{2} + \frac{1}{2} - 0 - 0 = 1,$$

$$\langle A_d B_b \rangle = (+1)(+1)P(+1,+1 \mid a',b) + (-1)(-1)P(-1,-1 \mid a',b)+$$
$$(+1)(-1)P(+1,-1 \mid a',b) + (-1)(+1)P(-1,+1 \mid a',b)$$
$$= \frac{1}{2} + \frac{1}{2} - 0 - 0 = 1,$$

$$\langle A_d B_c \rangle = (+1)(+1)P(+1,+1 \mid a',b') + (-1)(-1)P(-1,-1 \mid a',b')+$$
$$(+1)(-1)P(+1,-1 \mid a',b') + (-1)(+1)P(-1,+1 \mid a',b')$$
$$= 0 + 0 - \frac{1}{2} - \frac{1}{2} = -1,$$

which means $\gamma_{\exp} = 4$. This far exceeds the Tsirelson bound. So again, the question is, can we somehow physically instantiate the PR-box?

8.3 Quoin Mechanics

We ask because quantum information theorists quickly realized that if the joint probabilities of the PR-box could be instantiated physically, they would provide an 'unreasonably effective' means of communication. Let us follow Tanya and Jeffrey Bub's example of a "quantum guessing game" to illustrate this point [2].

To help visualize the PR-box, the Bubs introduced superquantum coins called *quoins*. If both members of a pair of entangled quoins are flipped starting as heads (HH), they will end up tails–heads $(-1,+1)$ or heads–tails $(+1,-1)$, never heads–heads $(+1,+1)$ or tails–tails $(-1,-1)$ ($\neq$ in Table 8.1). This corresponds to the last PR-box probability with $a' = b' = H$. Given any other starting configuration, i.e., tails–tails (TT) or tails–heads (TH) or heads–tails

(HT), they will end up "equal," i.e., heads–heads $(+1, +1)$ or tails–tails $(-1, -1)$ (Table 8.1). These are magical coins indeed, as the Bubs show their behavior cannot be explained by any "rigging" based on their starting positions. [Keep in mind that at this point we are simply trying to understand how each quoin might behave deterministically to explain the quoin mechanics of Table 8.1, not the probabilistic PR-box itself.]

To show this, they note that there are only four ways to rig a quoin:

1. No matter how it starts, it ends up heads (H).
2. No matter how it starts, it ends up tails (T).
3. It stays the same way it starts (S for same).
4. It changes from the way it starts (O for other).

They then list all the possible riggings for a heads–heads (HH) start that yield nonequal outcomes, and all the possible riggings for a heads–tails (HT) or tails–heads (TH) or tails–tails (TT) start that yield equal outcomes. As a result, it is apparent that there is no rigging that can yield the quoin mechanics of Table 8.1. Therefore, we simply have to accept quoin mechanics without knowing how they are instantiated in order to explore their implications for the Bubs' quantum guessing game.

Recall from Chapter 2 that Mermin showed something similar in trying to explain the quantum-mechanical joint probabilities for the spin triplet state using instruction sets. In that case for quantum joint probabilities, as with these superquantum joint probabilities, the instruction sets did not work and we were left with no constructive account of those quantum joint probabilities. Of course, the difference is that while we do not have any (consensus) constructive explanation for the quantum probabilities of the Bell states, we do instantiate them in the lab. And, as we just showed in Chapter 7, those quantum probabilities do obey a very reasonable conservation principle ('average-only' conservation per NPRF) while we will see that the PR-box probabilities violate

Table 8.1 Quoin mechanics. Results of flipping pairs of entangled quoins in four possible starting combinations.

		Alice's quoin	
		H	T
Bob's quoin	H	≠	=
	T	=	=

that principle. Now we show how these conservation-violating PR-box probabilities are 'miraculous' in their information exchanging capability. We start
with the game itself.

8.4 Tanya and Jeffrey Bub's Quantum Guessing Game

As you can see in Table 8.2, there are five lanes on each side of a barrier, which
does not allow Alice to see Bob's values and vice versa. The lanes are num-
bered 1 through 5 and the game is started by the dealer setting values of 1 or 0
for each player in each of the five lanes (Table 8.2). After the dealer has set these
ten values, we see that there are some lanes that have a 1 on both sides of the
barrier, e.g., lanes 1 and 4 in Table 8.2.

Alice is the "guesser" and because of the barrier she does not know what
Bob has in his lanes. Her job is to guess whether there are an even number of
lanes that have a 1 on both sides (as is the case in Table 8.2) or an odd number of such lanes. The only way she can know the answer with certainty is if
the dealer were to set all five of her lanes to 0, in which case she knows the
answer is even, i.e., there are zero lanes that have a 1 on both sides. Of course,
the dealer is not going to do that, so she will always have at least one 1 on her
side. For all such cases, it is not too difficult to convince yourself that the probability of the right answer being odd (even) is 50% (50%). Alice and Bob buy
six poker chips to play and the House doubles their chips for a correct guess.
[They turn their poker chips back into money when they leave the Quasino
of course.] If Alice guesses wrong, they lose all their chips to the House. Since
the odds are 50–50 of guessing right, Alice and Bob end up breaking even if
they continue to play, meaning they will not win or lose money overall on
average.

Table 8.2 An example game table. When Alice and Bob flip their
quoins for their 1 and 0 values shown here for each lane, together
they will have an even number of H in lanes 1 and 4 and an odd
number of H in lanes 2, 3, and 5 per quoin mechanics (Table 8.1).

Lane	1	2	3	4	5
Bob's values	1	0	1	1	0
		BARRIER			
Alice's values	1	0	0	1	1

But, there is another aspect of the game that we have not told you: Bob can spend one poker chip to send one bit of information to Alice. In that case, Alice and Bob have five chips remaining and, if they win, the House pays them five chips, i.e., the House doubles their remaining poker chips. Likewise, Bob can spend two chips to send two bits of information to Alice, which means the House would pay four chips if they win. Once they have to spend three chips to win, the House pays three chips and they break even monetarily by winning the game.

For example, Alice could ask Bob to send her the values in his lanes 1, 4, and 5 of Table 8.2, so she can see if he has any 1s in those lanes. That is all the information she needs, since she has 0s in lanes 2 and 3. Bob sends the bits 1, 1, and 0, which tells her they have just two lanes (lanes 1 and 4) where they both have a 1, so the answer is even. They win the game and three chips, but they spent three chips to send the three bits of information, so they broke even monetarily. You can convince yourself that this strategy would actually end up losing money in the long run.

Now it is time to show that if Alice and Bob use quoins, then they can win this game every time by merely passing one bit of information each game, thereby winning five chips every game, netting four chips every game. That is, they spend six chips to play and pass the one bit of information, while they receive ten chips in return for winning. Here is the Bubs' strategy.

Alice and Bob start with five pairs of quoins. [Do not confuse these with the six chips they bought from the House to play the game, they brought these five pairs of quoins with them to the Quasino.] Again, Alice is the guesser, so it is Bob who will be sending the one bit of information. Alice's quoins are labeled 1A, 2A, 3A, 4A, 5A, and Bob's are labeled 1B, 2B, 3B, 4B, 5B. Quoins 1A and 1B are entangled per quoin mechanics, as are 2A and 2B, etc. The number on each quoin corresponds to each lane of the game. For all lanes in which Alice has a 1, she flips the corresponding quoin starting with H. In all lanes with a 0, she flips the corresponding quoin starting with T. Bob does likewise with his quoins for his values and lanes. Bob then pays one chip to send Alice one bit of information: 1 if he has an odd number of Hs and 0 if he has an even number of Hs. From this one bit of information, Alice now knows with certainty whether they have an even or odd number of lanes with two 1s. The strategy is simple, although the reasoning behind it is not trivial.

The key is to observe that the individual outcomes of Alice and Bob's quoin tosses do not really matter, but their pairs do matter. In any lane with two 1s, Alice and Bob together observe a total of one H after they flip their quoins for that lane (quoin mechanics), which is odd. In the other lanes, they observe an even number (0 or 2) of Hs (quoin mechanics). Therefore, their *combined* count of Hs is even if and only if there are an even number of lanes with two 1s. Neither player by themselves knows whether the combined number of Hs is even or odd because each person can only see the outcomes of his/her own quoins. But, all Alice needs to know after flipping her five quoins is whether or not Bob has an even number of Hs or an odd number of Hs, and he can send her that one bit of information (1 for odd and 0 for even, for example) by spending just one chip. So, quoin mechanics guarantees they will win five chips every game (netting four).

8.5 Unreasonably Effective

How egregious is this advantage? In other words, can quantum joint proba-bilities achieve anywhere near this success rate? Let's look at the PR-box. We see that the fourth PR-box probability corresponds to the HH case produc-ing unequal outcomes in quoin mechanics. That is, using $a = b = $ T and $a' = b' = $ H with outcomes $+1 = $ H and $-1 = $ T, the PR-box aligns with quoin mechanics (Table 8.1). Since it is the HH case that allows us to discern even or odd pairs of 1s in our guessing game using only one bit of infor-mation, let's scrutinize the fourth PR-box probability using our conservation principle.

According to quantum mechanics, the joint probability of measuring like results for a triplet state in its symmetry plane is $\cos^2(\theta/2)$ and the joint prob-ability of measuring unlike results is $\sin^2(\theta/2)$, where θ is the angle between $\hat{a}$ and $\hat{b}$. The first PR-box probability says that $\hat{a} = \hat{b}$ (same results), the second PR-box probability says $\hat{a} = \hat{b}'$ (same results), and the third PR-box probabil-ity says $\hat{a}' = \hat{b}$ (same results). So, these three PR-box joint probabilities in total say $\hat{a} = \hat{a}' = \hat{b} = \hat{b}'$. That means we need the fourth PR-box probability to be in accord with $\hat{a}' = \hat{b}'$, i.e., predict the same results, but of course it says we must get opposite results, which means $\hat{a}' = -\hat{b}'$. Therefore, the joint probabil-ities of the PR-box violate our conservation principle in a maximal fashion for the triplet states (Table 8.3). [A similar argument can be made using the singlet state.]

Consequently, in order to satisfy our conservation principle we need the outcomes of an HH start to be equal, just as the outcomes of a TH, HT, or TT start are. But, if we make the outcomes of an HH start equal, we lose the advantage of quoin mechanics. In fact, using quantum coins (the outcomes of an HH start are equal) instead of superquantum quoins (the outcomes of

Table 8.3 Why the quantum? = Why the Tsirelson bound? [The "constraint" is conservation per no preferred reference frame.]

	CHSH quantity	
$-2 \leftrightarrow 2$	$-2\sqrt{2} \leftrightarrow 2\sqrt{2}$	PR correlations → 4
Satisfy Bell inequality	Tsirelson bound	No-signaling max
Classical correlations	Quantum correlations	Superquantum correlations
Violate constraint	Satisfy constraint	Violate constraint

an HH start are not equal) puts us right back to a non-quantum chance of winning the guessing game (recall the reasoning behind the quoin strategy).

So, we see that the superquantum PR-box is not just a little bit better than quantum mechanics for the quantum guessing game, it is 'unreasonably effective'. But, again, the PR-box violates our conservation principle, so it is probably the case that a physical instantiation of the PR-box is a pipe dream akin to a perpetual motion machine [8].

8.6 Summary

The no-signaling joint probabilities introduced by Popescu and Rohrlich (PR-box) violate the CHSH Bell inequality well beyond the Tsirelson bound. This stronger correlation enables the PR-box to win Tanya and Jeffrey Bub's "quantum guessing game" every time, while quantum mechanics cannot improve on guessing alone. So, if the PR-box could be physically instantiated, it would permit exceptional information transfer. Unfortunately, the PR-box violates conservation per NPRF in maximal fashion, so it is no more likely to exist physically than a perpetual motion machine.

To summarize the book to this point, a scientific paradox is resolved by refuting any of the paradoxical facts or by correcting the reigning scientific worldview so that the paradoxical facts make sense. We have resolved the EPR and EPR–Bell paradoxes by correcting the reigning scientific worldview, i.e., by prioritizing principle explanation over explanation via dynamical laws or mechanistic causal processes. Essentially, the mysterious correlations resulting from quantum entanglement are explained by Einstein's principle rather than Reichenbach's. In the next (and final) chapter, to borrow Bell's language from a Chapter 7 quote, we will speculate on the "radical conceptual renewal" [1] for the reigning scientific worldview that results from this correction to it.

References

[1] J. Bell, *Introductory Remarks*, Physics Reports, 137 (1986), pp. 7–9.

[2] T. Bub and J. Bub, *Totally Random: Why Nobody Understands Quantum Mechanics*, Princeton University Press, Princeton, NJ, 2018.

[3] B. S. Cirel'son, *Quantum generalizations of Bell's inequality*, Letters in Mathematical Physics, 4 (1980), pp. 93–100.

[4] J. F. Clauser, M. A. Horne, A. Shimony, and R. A. Holt, *Proposed Experiment to Test Local Hidden-Variable Theories*, Physical Review Letters, 23 (1969), pp. 880–884.

[5] B. Dakic and C. Brukner, *Quantum Theory and Beyond: Is Entanglement Special?*, in Deep Beauty: Understanding the Quantum World through Mathematical Innovation, H. Halvorson, ed., Cambridge University Press, New York, 2009, pp. 365–392.

[6] N. D. Mermin, *What's Wrong with this Criticism*, Foundations of Physics, 35 (2005), pp. 2073–2077.

[7] S. Popescu and D. Rohrlich, *Quantum nonlocality as an axiom*, Foundations of Physics, 24 (1994), pp. 379–385.

[8] W. M. Stuckey, M. Silberstein, T. McDevitt, and I. Kohler, *Why the Tsirelson Bound? Bub's Question and Fuchs' Desideratum*, Entropy, 21 (2019), p. 692.

9

What's Really Going On?

One way or another, God has played us a nasty trick. The voice of Nature has always been faint, but in this case it speaks in riddles and mumbles as well. Quantum theory and Relativity seem not to directly contradict one another, but neither can they be easily reconciled. Something has to give: either Relativity or some foundational element of our world-picture must be modified. Physicists may glory in the challenge of developing radical new theories in which nonlocality and relativistic spacetime structure can more happily co-exist. Metaphysicians may delight in the prospect of fundamentally new ontologies, and in the consequent testing and stretching of conceptual boundaries. But the real challenge falls to the theologians of physics, who must justify the ways of a Deity who is, if not evil, at least extremely mischievous.

Tim Maudlin (2011)

9.1 Without Malice

In our view, Maudlin's book is an argument for the conclusion in the opening quote of this chapter [23, p. 221], i.e., his book is the result of an extensive analysis of the implications of Bell's theorem for constructive accounts of the world. Essentially, Maudlin's book reveals that the violation of Bell inequalities tolls for constructive explanation at the most fundamental level. Luckily, there is a compelling alternative to constructive explanation, as we have shown in this book.

According to our proposed completion of the information-theoretic reconstruction program via principle explanation, the mystery of entanglement is solved by Information Invariance & Continuity as justified by the relativity principle (NPRF + h), so the Deity is neither evil nor mischievous, but benevolent in an impartial fashion. In short, NPRF + h reveals that 'God has made the world' so that quantum phenomena may be understood via

Einstein's Entanglement. W. M. Stuckey, Michael Silberstein, and Timothy McDevitt, Oxford University Press.
© Oxford University Press (2024). DOI: 10.1093/9780198919698.003.0010

empirical investigations from any and all reference frames of our data collection devices, because the mathematical regularities (including their constants of Nature) that provide an understanding of the data are valid in all of those reference frames. That is, NPRF + h dictates the kinematic structure of quantum mechanics (Hilbert space) exactly like NPRF + c dictates the kinematic structure of special relativity (Minkowski spacetime).

To this point in the book, we have intentionally invoked minimal ontological assumptions, choosing a phenomenological and methodological approach to creating an objective spacetime model of reality consistent with our principle solution to the mystery of entanglement. Since our solution to the greatest mystery in physics is a principle one, we had this luxury and did not want to unnecessarily alienate readers who were committed to a particular constructive ontology. Since any ontology that maps to our dynamical experience must contain or directly account for the corresponding constructive aspects of experience, we did not want to unnecessarily rule out how that might be done. Again, our principle solution, Information Invariance & Continuity as justified by the relativity principle (NPRF + h), is consistent with any number of constructive ontologies, such as particles, fields, or waves.

In this last chapter and its subchapters we speculate on the answer to Adam Becker's question, "What is Real?" beyond God's design principle, e.g., what does this principle range over or apply to? It is important to understand that while our account is a principle one and a psi-epistemic one (Subchapter 9A), it is perfectly compatible with scientific realism[1] and a constructive ontology such as particles and fields. As this chapter will demonstrate, one can have a principle explanation for entanglement that ranges over a constructive ontology so as to avoid instrumentalism or operationalism (defined in Chapter 1). That is, a principle account need not dodge the question of what entities are fundamental in physics, as we will show using an ontology based on multiscale contextual emergence in 4D spacetime (explained below) and the Poincaré symmetries. We should warn the expert reader that this chapter is merely a sketch of an ontology, a detailed ontology would require a book in itself, e.g., our book, "Beyond the Dynamical Universe" [34]. [Note: We will occasionally reference Subchapters 9A–9D that immediately follow the main chapter and contain digressions that are set aside so as not to interrupt the flow of the main chapter.]

---

[1] According to entity scientific realism, the theoretical entities such as electrons and photons mentioned in our scientific theories are real.

9.2 Contextuality Versus Reductionism and Relationalism

The problem with constructive interpretations of quantum mechanics is that they are stuck with nonlocality, violations of statistical independence, etc., but a virtue of such models is that they are realist in nature, attempting to tell us about the nature of reality. A worry is that any principle and psi-epistemic account of the quantum will fail to be a realist one (Subchapter 9A). In this chapter we want to provide our realist story about the nature of reality and show how it, along with our principle explanatory schema, resolves many explanatory and ontological conundrums explicitly involving quantum entanglement or superposition, such as the double-slit, delayed-choice, Schrödinger's Cat, and Wigner's Friend experiments (Subchapters 9C and 9D). We will also briefly explain how our all-at-once perspective might advance quantum gravity (the attempt to unify quantum physics and general relativity), avoid singularities in general relativity, and resolve the mysteries of dark matter and dark energy (Subchapter 9B). Our 4D all-at-once model of reality will certainly contain some twists relative to realist models that assume that the existence and behavior of classical entities are best explained constructively.

We think that quantum mechanics and relativity are telling us that the universe is fundamentally 4D (in that 4D adynamical or acausal global constraints (AGCs) are fundamental to constructive causal patterns and dynamical laws) and, as suggested by Kochen and Specker, and other theorems and experiments [22], inherently contextual in nature (in that the 4D constraints range over 'exchanges' of quanta in classical contexts). We call this *quantum–classical contextuality*, which is a subset of what we call *multiscale contextual emergence*. Such a world is not inherently compositional in nature, i.e., it violates what Weinberg calls "constituent reductionism" or "petty reductionism." He dismisses petty reductionism as follows [35]:

> Petty reductionism is not worth a fierce defense. Sometimes things can be explained by studying their constituents—sometimes not ... In fact, petty reductionism in physics has probably run its course ... It is also not possible to give a precise meaning to statements about particles being composed of other particles. We do speak loosely of a proton as being composed of three quarks, but if you look very closely at a quark you will find it surrounded with a cloud of quarks and antiquarks and other particles, occasionally bound into protons.

But from our perspective, Weinberg does not go far enough, because he still presupposes that the arrow of explanation will always be bottom-up, from smaller scales to larger ones, and that fundamental explanation will be constructive. We believe relativity and quantum mechanics are telling us otherwise. We call this multiscale contextual emergence because it suggests that determination relations between smaller and larger scales can be from smaller to larger or vice versa. Contextual emergence emphasizes the ontological and explanatory fundamentality of 4D and other global constraints operating over entities at multiple scales whose existence and properties are inextricably interdependent. This is a unifying fact about the nature of reality. [For a book-length treatment of this idea see [6].] We argue that quantum mechanics exhibits a specific kind of multiscale contextual emergence we call quantum–classical contextuality.

In our model of reality, there can be no classical objects without exchanges of quanta, and vice versa. Given contextual emergence, sometimes states of affairs at larger scales provide necessary conditions for the existence of micro-entities and their properties at smaller scales. And states of affairs at smaller scales sometimes provide only necessary and not sufficient conditions for states of affairs at higher scales. Do not worry, we will provide examples to clarify this idea.

Of course, there is some sense in which classical objects are made of atoms, but as we will show, whether or not an atom behaves in a classical or a quantum fashion is a function of the 4D quantum–classical context. And this is not just true for atoms, but for everything! So, the existence and behavior of all things does not merely supervene on their parts and their properties at a particular time, but on the 4D 'exchanges' and 'interactions' of quanta with other things.

If one is a petty reductionist and thus assumes constructive explanation is fundamental, this is a crazy thing to think. Indeed, if one does not pay attention, it may sound like instrumentalism or something Neils Bohr would say. But, the co-dependence and co-determination of the quantum and the classical as given by global 4D constraints ranging over what is truly a 4D universe is another story. All of this will be explained in this chapter. Here, one needs to think of the world not as a chess game or finite automata like Conway's game of life, but as a crossword puzzle with the words (4D distribution of mass–energy in spacetime) filled in self-consistently as given by the clues (the 4D adynamical/acausal global constraints).

9.2.1 Comparison with Relational Quantum Mechanics

As we noted earlier, there have been a few other attempts to provide all-at-once accounts of the quantum, e.g., Ghirardi, Rimini, and Weber (GRW) flash theory [15] and Rovelli's relational quantum mechanics [1, 2]. Once again, it turns out that our account of the quantum has the most affinities with Adlam and Rovelli's relational quantum mechanics. Take, for example, the very analogy that launched relational quantum mechanics [28, p. 135]:

> Einstein extended the notion of relativity to time: we can say that two events are simultaneous, only relative to a given motion ... Quantum mechanics extends this relativity in a radical way: all variable aspects of an object exist only in relation to other objects. It is only in interactions that nature draws the world.

Of course, unlike Rovelli's relational quantum mechanics, our view does not make the outcomes of quantum experiments relative to "observers," but rather we show how NPRF + h is analogous to NPRF + c. That is, just as with special relativity, it is not the relativity part that is the deepest fact, it's the invariance. This difference in how to explain or view quantum entanglement turns out to be very important for a number of reasons.

For one, as we will discuss below, Rovelli's original version of relationalism leads to a serious problem for a realist account of the quantum involving failures of intersubjective agreement, sometimes called the Wigner's Friend problem (Subchapter 9D). Needless to say, any account of the quantum that entailed an unresolvable failure of intersubjective agreement of the sort that drives Einstein's relativity would entail a malicious world indeed. But, as we already noted, in Adlam and Rovelli's relational quantum mechanics [1] there is an attempt to resolve this problem and move relational quantum mechanics in an all-at-once direction.

Although there are some very important differences, Rovelli's relationalism is very much in the spirit of our emphasis on contextuality. As he says, "The events of nature are always interactions. All events of a system occur in relation to another system" [28, p. 137]. Even more tellingly, Rovelli argues for "the impossibility of separating the properties of an object from the interactions in which these properties manifest themselves and the objects to which they are manifested" [29, pp. 78–79]. In much the same fashion as for our contextual emergence, he goes on to say:

> The properties of an object are the way in which it acts upon other objects; reality is this web of interactions ... for there are no properties outside of interactions ... When the electron does not interact with anything, it has no physical properties. It has no position; it has no velocity.

We agree completely with the spirit of these remarks, but again, as we will make clear in this chapter, there are important differences between Rovelli's relationalism and our contextuality. One big difference will turn out to be the implications of quantum–classical contextuality. Relational quantum mechanics starts with an analogy from special relativity, but the universal relationalism is all quantum in Rovelli's view. For example, regarding the original version of relational quantum mechanics, Adlam and Rovelli note [1]:

> We reinforce that in this picture, the variables of a system take on values only during a quantum event, and at all other times they have no values. This is the way in which [relational quantum mechanics] makes sense of the Kochen–Specker theorem and other contextuality theorems.

From the very beginning, our principle-based explanatory schema and our all-at-once and contextual model of reality have been inspired by Adlam and Rovelli's relational quantum mechanics. This is no doubt apparent to any reader who has made it this far in the book and it will become even more apparent moving forward in this last chapter. To move forward in this chapter, the reader interested in what is required for a realist account of quantum mechanics and how our account of quantum mechanics is realist should next read Subchapter 9A. If you are not interested in that digression, you can continue to Section 9.3 in this main chapter.

9.3 What is Real?

In his 2018 book *What is Real? The Unfinished Quest for the Meaning of Quantum Physics*, Becker writes [5, p. 7]:

> science is about more than mathematics and predictions—it's about building a picture of the way nature works. And that picture, that story about the world, informs both the day-to-day practice of science and the future development of scientific theories, not to mention the wider world of human activity outside of science.

Becker's "picture of the way nature works" is what Maudlin called a world-picture or worldview, and it is what we have been calling the objective spacetime model of reality per the game of physics, i.e., a model of reality resulting from intersubjective agreement about our empirical observations per Zeilinger. Recall, physicists are trying to discover patterns in the relationships between, and events involving, the bodily objects of their empirical investigations, so as to unify what are otherwise quite different data in one spacetime model of the "real external world," i.e., an objective spacetime model of reality. Per Einstein [13]:

> ... the totality of our sense experiences is such that by means of thinking (operations with concepts, and the creation and use of definite functional relations between them, and the coordination of sense experiences to these concepts) it can be put in order ...

What we have seen throughout physics is that the fundamental principle reconciling the otherwise disparate collection of empirical data is that:

No empirical data provide a privileged perspective on the real external world

So, our principle approach to physics shows that physics is comprehensively coherent despite objections otherwise by Einstein, Bell, Dirac, and others. While physics is not finished, as we will show below, quantum information theorists have revealed a tremendous underlying unity in the physics we do have.

The history of the relativity principle from Galileo and Newton to Einstein to Rovelli, Zeilinger, Brukner, Fuchs, Hardy, Bub, Clifton, Halvorson, Grinbaum, Dakić, Masanes, Müller, Höhn, et al. culminates with Information Invariance & Continuity as justified by NPRF. Again and again, Nature's adherence to the relativity principle has forced us into "radical conceptual renewal."

With Galileo and Newton we had to abandon a purpose behind forces. Unlike the Aristotelian physics it replaced, Newtonian mechanics provides a mathematical formula for how forces act on bodily objects, but it does not provide any reason why they act.

With Einstein we had to abandon a fundamental constructive account. Special relativity adheres to a fundamental causal structure, but there is no dynamical law or mechanistic causal process behind its fundamental principle of NPRF + c.

Finally, with quantum mechanics we have had to abandon cherished notions about causality. Specifically, the violation of Bell's inequality tells us to abandon Reichenbach's Principle, i.e., that correlations must be explained in conventional causal fashion, at least at the most fundamental level.

In fact, quantum information scientists are already exploiting this feature of quantum mechanics. For example, Mateus Araújo, Veronika Baumann, Časlav Brukner, Fabio Costa, Adrien Feix, Philippe Allard Guérin, Marius Krumm, Lorenzo M. Procopio, Lee A. Rozema, Giulia Rubino, Philip Walther, and Jonas M. Zeuner [3, 4, 30] have already shown that [11] "quantum circuits with indefinite causal ordering perform certain calculations faster than those with definite causal ordering." As John Preskill said in 2020 [27]:

> What I think quantum information science is about is we're in the early stages of the exploration of a new frontier of the physical sciences, sometimes I call it the "entanglement frontier" ... Many exciting surprises await us as we explore the entanglement frontier.

Indeed, Giulio Chiribella, Giacomo Mauro D'Ariano, Paolo Perinotti, and Benoit Valiron [8] designed a quantum switch based on indefinite causal ordering that "produces an output circuit where the order of the connections is controlled by a quantum bit, which becomes entangled with the circuit structure."

In his March 2021 *Physics Today* editorial [12], "Quantum information is exciting and important," Charles Day mentions "entanglement-based clocks of unprecedented precision." He also points out that a "76-qubit device based on entangled photons solved a sampling problem 10^{14} times faster than a classical device could." With such profound technological implications, the United States Department of Energy and its private sector partners provided $1 billion in 2020 to fund Quantum Information Science Centers [14].

Again, we are arguing that in some cases it is preferable to abandon causal talk altogether. So, let's see what we might infer about reality from the principle approach to physics outlined in this book.

9.3.1 Reality is 4D

From the relativity of simultaneity, our objective spacetime model of reality is four-dimensional or all-at-once. Again, as we pointed out in Chapters 1, 3, and 4, each person's subjective spacetime model of reality is dynamical and contains an unambiguous notion of what is happening everywhere in space right NOW. For example, if NASA scientists are sending signals to a probe en route to Jupiter, they need to know where that probe is NOW and how it is moving NOW. Then they can calculate where it will be and how it will be moving when their signal reaches it 45 minutes later. We can communicate with each other without confusion as long as our subjective spacetime models of reality have overlapping NOW slices, e.g., your friend texts you that a particular grocery store is open NOW, but closes in 15 minutes. That agreement about what is happening NOW was not true for the girls and boys in Chapter 4.

As we showed in Chapter 4, according to the shared NOW slices for sisters Alice, Kim, and Sara, they were all the same age while Bob is two years older than his brother Joe. Conversely, according to the shared NOW slices for Bob and Joe, they were the same age while Kim is 1.6 years younger than Sara and Alice is 2.5 years younger than Sara. All of this follows from NPRF and the fact that the speed of light c (establishing a limit on the speed of information transfer) is not infinite.

Consistent with the fact that our objective spacetime model of reality is all-at-once per the relativity of simultaneity is the fact that our second universal constant of Nature, Planck's constant h, is a unit of action. As we pointed out in Subsection 1.6.1, action is angular momentum multiplied by angular distance or energy multiplied by time or momentum multiplied by distance, and then added up along the worldline for a particle. Let's pause to explain that.

As Planck and Einstein showed, the energy E of a photon is given by Planck's constant h and the frequency f of the photon, i.e., $E = hf$ (the Planck–Einstein relation). If one thinks of the frequency of the photon as the inverse time of emission or absorption of the photon $f = 1/T$, then we have $E \cdot T = h$. Keep in mind that what is really true about the photon is based on its properties and those are defined by the 4D context. And, we know it is impossible to establish a single context in which both energy and time are determined exactly for a quantum because they are complementary properties. However, as we will see in Subsection 9.3.2, properties and therefore 'things' are contextually defined so they are not as localized in 4D as we think of them.

With that caveat, let's suppose a photon is emitted upwards from the surface of Earth in a time interval T_e with energy E_e and is absorbed at a higher elevation in a time interval T_a with energy E_a. Since the photon loses internal energy to gravitational potential energy, $E_a < E_e$. Therefore, the photon is absorbed over a longer time interval $T_a > T_e$, i.e., the frequency of the photon is reduced, i.e., its wavelength is increased in what is called *gravitational redshift*. Then we have $E_e T_e = E_a T_a = h$ due to the invariance of the elementary unit of action h.

Likewise, a massive particle can be assigned a wavelength $\lambda = h/p$ (called its *de Broglie wavelength*) based on its momentum p, just like a photon. [$E = pc$ in special relativity for massless particles, $E = hf$ in quantum mechanics for photons, and $c = f\lambda$ in wave mechanics, so we have $p = h/\lambda$ for photons.]

If one thinks of the wavelength λ of the massive particle as the spatial length over which emission or absorption occurs, then we have $p \cdot \lambda = h$ for the massive particle just like $E \cdot T = h$ for the massless photon. So, if the emitted momentum p_e is greater (less) than the absorbed momentum p_a, then the length of emission λ_e is less (greater) than the length of absorption λ_a. Again we have $p_e\lambda_e = p_a\lambda_a = h$ due to the invariance of the elementary unit of action h. All of this follows from NPRF and the fact that Planck's constant (establishing a limit on the amount of simultaneous information available) is not zero.

So, as discussed, the "picture of the way nature works" is that bodily objects interact by exchanging quanta of energy and momentum in (nearly) discrete emission and absorption events of the same action h (with the caveat above). What are these quanta? Well, if you attempt to describe them as bodily objects or classical fields existing between emission and absorption events in space-time, i.e., you attempt a constructive account of quantum mechanics, you run into all the problems we pointed out in the book. That is, no matter how you try to characterize them constructively you will end up with entangled quanta that violate locality, statistical independence, etc.

As with special relativity, there may never be any consensus constructive account for the exchange of quanta, although this is precisely what Wharton and Argaman are working on for retrocausality and Hance, Hossenfelder, and Palmer are working on for superdeterminism, as we pointed out earlier. In both programs, a successful constructive account promises a completion of quantum mechanics with new experimental predictions, so these efforts should be pursued even if at present the prognosis is not good. Since we are not exploring those constructive options, let's rather explore in more detail the ontology we infer from the principle account of quantum mechanics explained in this book.

9.3.2 Quanta and Bodily Objects

Again, we are talking about the exchange of quanta between the bodily or classical objects of our empirical investigations in accord with NPRF so that there is no violation of locality, statistical independence, intersubjective agreement, or unique experimental outcomes. That means the quanta do not possess instruction sets / hidden variables / counterfactual definiteness, so we cannot think of quanta in terms of classical physics. Since 'things' (ontic entities) are defined by their properties, we need to be clear as to how our properties are defined. In our case, we are just using the properties defined in physics, e.g., mass, position, velocity, momentum, energy, spin, polarization, ...

Physicists define properties so as to facilitate formal relationships, e.g., the ideal gas law $PV = NkT$, so these properties are often ascertained/measured in highly contrived/controlled fashion. That is, any measurement is simply a series of contrived and recorded interactions and outcomes in the context of bodily objects distributed in space and time for the purpose of isolating the values of specific properties according to physics. So, to understand the properties defining ontic entities it is necessary to refer to measurements. However, these properties exist for bodily objects whether or not there are physicists controlling the contexts and noting the outcomes of the interactions, since the properties of bodily objects commute. That is, bodily objects possess instruction sets / hidden variables / counterfactual definiteness, as we explain below. With that caveat, let's look more closely at the relationship between bodily objects, classical fields, and quanta.

Interactions and Bodily Objects

What exactly do we mean by a *bodily object*? As we quoted Einstein in Chapter 1, our observations involve interacting bodily objects understood via "repeatedly occurring complexes of sense impression." [Of course, we are not claiming classical objects are nothing but sense impressions and neither was Einstein.] For example, you see a pen on your desk, you look away and when you return your gaze to your desk, you again see a pen at rest with respect to you and the desk. You associate the first sense impression of a pen with the second. Indeed, even if you do not change your gaze, the pen persists through the temporal evolution of your sense impressions in that same location on the desk. We would say that it has a worldline (or worldtube if its spatial extent is also represented) in your subjective spacetime model of reality where you, the pen, and the desk share an inertial reference frame.

So, important properties of a bodily object are its position and momentum, which we can know simultaneously. Recall, the variables for bodily objects in classical physics commute while the variables for quanta in quantum mechanics may not. In this case, position and momentum are complementary variables for quanta, so one cannot know both of them simultaneously for quanta. But, how are you obtaining this commuting information about the pen and the desk?

You are interacting with them via the exchange of quanta for which these properties do not commute. How can that be? Obviously, you (or your data collection devices) are exchanging lots and lots of quanta with those bodily objects. Some of those quanta contribute to position data while others contribute to momentum data. How many interactions are needed to establish the existence of a bodily object?

Clearly, the more quanta you can acquire from your source per unit time, the better you will be able to statistically establish that your source is a bodily object. For example, suppose you want to create a bodily object for a single atom. Recall, single silver atoms exchanged between the oven and detector in the Stern–Gerlach experiment were not bodily objects, they were quanta. In that experiment, the property of spin angular momentum was being measured and it did not commute in orthogonal directions for the quanta. But, we were only obtaining a single qubit of information on each atom and that is clearly not enough to establish a bodily object. So again, how many interactions are needed?

There is no hard and fast answer to that question because it is context dependent and necessarily statistical in nature, but let's look at how many interactions per second physicists used to establish the existence of a rubidium atom in an atomic trap [9] (a bodily object). In that experiment, physicists used light emitted from the atom to establish a worldline for "a rubidium atom stuck in an atomic trap." In doing so, the physicists were able to turn what was a quantum in the SG experiment (an atom with noncommuting angular momenta) into a bodily object (an atom with commuting position and momentum) for "several seconds." How many quanta of light were emitted by the atom to turn it into a bodily object? In that experiment, about 100 000 photons per second were used to create the bodily object we describe as "an atom stuck in a trap" (commuting position and momentum) like the pen on your desk.

Multiscale Contextual Emergence

As we alluded to earlier with regards to Rovelli's relationalism and our multiscale contextual emergence, the atom in these examples is not an underlying bare 'thing' (beable) that acquires properties in various contexts. A bodily object is a worldtube in spacetime defined by the various 4D interaction contexts for that worldtube (whether they are measurements or not). A quantum is an action h in the worldtube of a bodily object defined by properties that are themselves defined by a specific 4D interaction context of bodily objects, e.g., beam splitters, polarizers, detectors, sources, mirrors, magnets, etc., as we have seen. We talk as if we have a beable that is a quantum in one 4D context and a bodily object in another, because we have a source that is a bodily object whose worldtube links those different 4D contexts, and since the source is a bodily object, counterfactual definiteness holds. This is necessary for reasonable communication, e.g., I want to be able to say, "Consider a source of alpha particles in the following experiment (4D context)" without having to articulate all the 4D contexts necessary to define that "source of alpha particles." But, let's pause to do just that.

There are many different contexts that can be used to determine we have a source of alpha 'particles' with momentum p (note the scare quotes, we will return to this point later). We could put the source in a cloud chamber and use successive, colinear clicks to establish a velocity v. Then we could make a charge measurement using induced voltage from the beam of particles, which we subsequently count as a function of time via impingement on a detector to get the charge q per particle. Next we could send the particles through a double-slit screen onto a detector and get the momentum p using λ from the resulting interference pattern, i.e., $p = h/\lambda$ (Subchapter 9C). Or, we could find the cutoff voltage required to stop the detection of particles coming from the source (like that in the photoelectric effect experiment) to get a measurement of momentum for the particles. Since $p = mv$ and we have v from our first 4D context, we now have the mass m. Altogether, these 4D contexts establish (statistically speaking) m, q, and v of the alpha 'particles' coming from the source.

We can check all of this against a 4D configuration of the cloud chamber with a known magnetic field intensity B directed perpendicular to the plane of motion. The resulting radius of curvature r of the detector clicks (water vapor droplets in the cloud chamber or activated CCD pixels in a solid-state detector) should be given by $r = mv/qB$. Indeed, this is how particle physics is done, i.e., one uses classical physics to establish the particle types in the data of a high-energy "event" and quantum field theory to predict the distribution of particle types (Section 9.4 and Figure 9.6).

So, where are the quanta in this example? Well, you have quanta in the context of the double-slit experiment (thus the scare quotes earlier for alpha 'particles'). In fact, since your 4D context allowed you to determine p for the quanta being emitted by the source, there was no information about position at all in the double-slit context (Subchapter 9C). That means the double-slit experiment did not establish v, so we cannot get m from that context either. Then do we really have "alpha particles" in that double-slit experiment? No, the required properties for an alpha particle are not established by that interaction context alone, and without all the defining properties we do not have the corresponding entity. Further, since the quanta in the double-slit experiment are not directly 'connected' to the quanta in any of the other 4D-property-defining contexts for our source (in accordance with no counterfactual definiteness/noncommutativity), except indirectly via the worldtube of the source, there is no way to assign or attribute all the necessary properties for alpha particles to the specific quanta in any one context. This is not an epistemological limitation, it is an ontological fact.

Where exactly in spacetime is our *source* of alpha particles? Everywhere along the worldtube of the source, as explained above. Where exactly are the *alpha particles*? The alpha particles are defined by the entire collection of quanta in all the aforementioned 4D contexts, i.e., the quantum events precipitating the worldlines of clicks in the naked cloud chamber, cloud chamber with B, detector for q measurement, detector for stopping potential configuration, and interference pattern of double-slit configuration. So, the alpha particles are no more localized in space at any given time than the source. Every 'thing' is 'there' in spacetime, but not 'there' in space at a given time. When you look at one particular 4D context, you do not have alpha particles, so you do not have a source of alpha particles. The alpha particles only exist in the larger 4D collection of those smaller 4D contexts.

In practice, every source is not subjected to every 4D context needed to establish all of its defining properties. It is usually assumed that a source manufactured in the same way as a tested counterpart is ontologically equivalent. We would say that the success of such pragmatism reflects the truth of the assumed ontological equivalency. So, sources that are not even used for experimental science, which constitute virtually all sources in Nature, are ontologically well defined in this counterfactual sense (keeping in mind noncommutativity for quanta, as explained above). That means we do not need a specific worldline to connect a collection of 4D contexts to establish ontological equivalency, as with our alpha particle example. Rather, far weaker associations, such as "manufactured in the same way," suffice to establish ontological equivalency between otherwise disjoint local 4D contexts in spacetime globally. The properties needed for ontological equivalency are not (necessarily) established in any one local 4D context, but across various local 4D contexts on various scales. All of this is *multiscale contextual emergence* [6]; nonetheless, everyone agrees about what exists throughout our objective spacetime model of reality, i.e., the ontology is totally objective, and there is nothing 'missing' or 'hidden.'

Photons and the Electromagnetic Field

As for the relationship between photons (quanta of electromagnetic energy–momentum exchange) and the electromagnetic field, look again at the example in Chapter 7. If vertically polarized light is passed through a polarizer angled at 45° with respect to the vertical, classical electromagnetism says half the light will pass. That is the classical property defining the polarizer (a bodily object). But, when we send vertically polarized photons to the polarizer, it is not the case that half of each photon passes through the polarizer; instead, half the photons pass and half do not, in violation of the classical property defining the polarizer. That is because 'half photons' passing through the polarizer in accord with classical physics would have their energy on the other side given by $E = h/2f$ instead of $E = hf$, meaning the value of Planck's constant was effectively cut in half at that polarizer orientation [39].

Notice that the vertically polarized state $|V\rangle$ can be understood as a superposition of diagonal and antidiagonal states (Figure 9.1):

$$|V\rangle = \frac{|D\rangle + |AD\rangle}{\sqrt{2}},$$

where

$$|D\rangle = \frac{|V\rangle + |H\rangle}{\sqrt{2}}, \qquad |AD\rangle = \frac{|V\rangle - |H\rangle}{\sqrt{2}}.$$

Accordingly, a vertical polarization measurement produces a V outcome with probability 1:

$$\left|\langle V|\frac{|D\rangle + |AD\rangle}{\sqrt{2}}\right|^2 = 1,$$

so all of the energy of the electromagnetic field is passed through the polarizer when every photon in the field is in this superposition state. If the polarizer

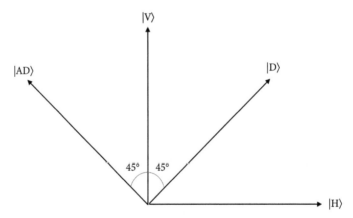

Fig. 9.1 Complementary Hilbert space qubits for photon polarization. The basis for a vertical measurement is $|V\rangle$ for vertical "pass" and $|H\rangle$ for horizontal "pass" (which is vertical "no pass"). The basis for a diagonal measurement is $|D\rangle$ for diagonal "pass" and $|AD\rangle$ for antidiagonal "pass" (which is diagonal "no pass"). These are complementary measurements for photon polarization, and the angle between the two bases is 45° just like the bases for the complementary spins S_z and S_x (Figure 6.4).

is rotated from vertical to diagonal or antidiagonal, the energy passed in each case is half. So, if the answer to the question "Is this photon vertically polarized in this state?" is "yes," then the answer to the question "Is this photon diagonally polarized in this state?" is totally ambiguous at 50% "yes" and 50% "no." That means the measurement for the $|V\rangle$ and $|H\rangle$ states is complementary to the measurement for the $|D\rangle$ and $|AD\rangle$ states (Figure 9.1).

However, this is not the same as saying the vertically polarized electromagnetic field can be thought of as a collection of photons where half the photons are diagonally polarized and half are antidiagonally polarized. That is called a *mixed state* (Chapter 6) as opposed to a superposition. If you conducted a vertical polarization measurement on such a mixed state, each diagonal or antidiagonal photon has a probability of 0.5 of passing the polarizer:

$$\left| \langle V | \frac{|V\rangle \pm |H\rangle}{\sqrt{2}} \right|^2 = \frac{1}{2},$$

so the energy of this electromagnetic field would be cut in half by a vertical polarizer. Indeed, a polarization measurement along any axis cuts the energy of this mixed state in half, so this mixed state corresponds to an unpolarized electromagnetic field. This is how we can model the classical electromagnetic field per compositions of quanta in superpositions and/or mixed states.

Summary
Putting all of this together, our world-picture is one of bodily objects that interact via the exchange of quanta. All of the properties of bodily objects commute, while not all of those same properties in quanta commute. Nonetheless, the properties of a bodily object are established via its interactions with other bodily objects. That means we need lots of interactions between our data collection devices and any given bodily object to establish its properties. And, when dealing with the exchange of large numbers of quanta, one can talk about their average, aggregate properties in terms of classical fields [31].

So, in the SG experiment, the classical magnetic field of the SG magnets is established by a large number of coherently organized (coaligned) magnetic moments of spin-$\frac{1}{2}$ particles. The quanta of silver atoms (defined via multiscale contextual emergence) constitute quanta of energy–momentum between the source (oven with silver gas) and the detector capable of rendering a photographic image of a location in space in response to silver atoms. That photographic image (worldline beginning at a quantum event) constitutes a value of spin angular momentum for the quantum of $\pm\hbar/2$ with respect to the direction of that SG magnetic field and with respect to that source.

Are we merely measuring an intrinsic pre-existing value of spin angular momentum for the quanta along the SG magnetic field with this experimental configuration? No, we have opted for an objective spacetime model of reality that does not violate locality, statistical independence, intersubjective agreement, or unique experimental outcomes, so we cannot have a constructive account of the quantum per Bell's theorem. Indeed, as we pointed out concerning a "source of alpha particles," the source and the alpha particles are not even localized to a particular spatiotemporal region. Rather, they are defined by multiscale contextual emergence, so we have to think about the properties of bodily objects and quanta differently.

9.3.3   Quantum–Classical Contextuality

Since the quantum being emitted by the source does not possess pre-existing, intrinsic values of its variables (instruction sets or counterfactual definiteness) independent of measurement (or interaction more generally), the values of $\pm\hbar/2$ (± 1 for short) in the SG experiment must be determined globally according to a specific SG magnet orientation in each trial of the experiment. For example, the value of $+1$ or -1 for the spin angular momentum of a silver atom in the SG experiment is not determined by the source alone, but by the oven with gaseous silver, a hole in that oven, a collimator, the SG magnet orientation, and a detector for silver atoms with its resulting image, all of which must be properly placed relative to each other in space. That is, the hole in the oven needs to be aligned with the collimator, then with the SG magnets, and that has to be aligned with the detector. And, all of those bodily objects are established by properties defined via 4D contexts along their worldtubes. Again, we have multiscale contextual emergence.

As we explained in Chapter 6, this is what motivates the information-theoretic axioms of relational quantum mechanics. And, again, no counterfactual definiteness per Rovelli's Axioms 1 and 2 is equivalent to Brukner and Zeilinger's "Information Invariance." "Continuity" requires an additional axiom in the reconstruction of relational quantum mechanics, as we pointed out in Chapter 6. Of course, Information Invariance & Continuity is justified by NPRF, so we see how relational quantum mechanics is based on NPRF precisely per Rovelli's desideratum. That is what we meant in Chapter 1 when we said that relational quantum mechanics is based on Information Invariance & Continuity, albeit in a different form.

That the values of various properties are context dependent is not unique to quanta, it is also true of ordinary bodily objects as we showed for the source in a "source of alpha particles." Both bodily objects and quanta are elements of the ontology, i.e., they are both things that exist, they simply differ in terms of what we might think of as their defining contextual information.

So, the properties defining bodily objects are measured/ascertained via the exchange of quanta between configurations of other bodily objects, and the properties defining quanta are also determined by experimental configurations of bodily objects such that the properties of bodily objects obtain on average from the properties of quanta. We call this form of multiscale contextual emergence *quantum–classical contextuality* and it simply means that because of the co-determination relations already discussed, bodily objects and quanta are defined using bodily objects and quanta in self-consistent, self-referential fashion. Let's revisit the simple example of photons and polarizers above.

A polarizer for electromagnetic radiation is a bodily object, and its property of interest is defined by its action on electromagnetic waves (classical field). As we explained in Chapter 7, the axis along which the electric field $\vec{E}$ oscillates in the electromagnetic wave is the wave's polarization, and a polarizer has a specific axis along which the electric field is allowed to oscillate without interacting with the polarizer (Figure 7.7). That is how the polarizing property of the polarizer (property of the bodily object) is determined using other polarizers and the electromagnetic field, i.e., in the context of other bodily objects and a classical field.

Now consider reducing the intensity of the incident wave until only one quantum of electromagnetic energy (photon) is hitting the polarizer at a time. If the photon (quantum) is transmitted by the polarizer (bodily object), the photon is polarized in the direction of the polarizing axis. That is how the polarization property of the photon (quantum of classical field) is determined by the polarizer (bodily object).

Sending these photons to the subsequent polarizer we find that each photon either passes or does not in accord with the Planck postulate and the Planck–Einstein relation as pointed out in Subsection 9.3.2. Quantum mechanics says the probability that each photon passes is $cos^2(\phi)$, so we see that the polarization property of the polarizer (property of the bodily object) obtains on average from the polarization property of the photon (property of the quantum), as determined by the polarizer (bodily object), in self-consistent, self-referential fashion. Again, we call this quantum–classical contextuality, and it turns out that contextuality in general is ubiquitous [6].

In quantum information theory, for example, whether or not a photon is a qubit with its two-dimensional Hilbert space depends on the context. In the polarization measurements of a photon above, the photon was a qubit and, as we will see in Subchapter 9C, a photon in the double-slit experiment is a qubit while in the triple-slit experiment the photon has a three-dimensional Hilbert space comprised of qubits. In fact, as you carve more and more slits into the screen within a finite width, i.e., going from triple-slit to quadruple-slit to quintuple-slit to … so as to increase the number of slits per unit length, the dimension of the corresponding Hilbert space grows and the complexity of qubit combinations increases. In the limit as the number of slits goes to infinity, you must recover the unobstructed wavefront, which is Huygens' principle.

So, we see how the wave of a classical electromagnetic field is an approximation of many, many photonic qubits per the Hilbert space structure. Notice that nothing is changing with the source of photons per se as we change the number of slits, yet our information exchange is changed dramatically from a qubit to a classical electromagnetic wave.

And while all this is beyond the scope of our book, in our view, environmental decoherence, so-called quantum nonseparability, quantum holism, quantum relationalism, quantum dispositionalism, etc., are really just symptomatic of the fundamentality of multiscale contextual emergence. And furthermore, while the contextuality in question is often manifested in dynamical and causal interactions, the deeper contextuality that explains and underwrites certain aspects of those interactions can be non-causal, non-dynamical, and spatiotemporal—what we have been calling 4D adynamical or acausal global constraints (AGCs). For example, as we demonstrated herein, the kind of contextuality we see in the case of EPR correlations is a consequence of NPRF. We see nothing inherently anti-realist about this view, as again, conservation laws and multiscale contextual emergence are real mind and perspective-independent facts about the world.

Integral Rather Than Differential Formalism

All of this means that quantum–classical contextuality is spatiotemporal/4D/all-at-once, meaning the properties of bodily objects and quanta are determined by interactions in the configurations of other bodily objects (polarizers, beam splitters, SG magnets, cameras, meter sticks, clocks, etc.) in some region of space through some interval of time. That is why the path integral formalism of quantum mechanics (mentioned in Chapter 3) more accurately aligns with the all-at-once ontology than the differential equation formalism (Schrödinger's equation). In the path integral approach, one computes the probability for each possibility directly without using a time-evolved wavefunction.

That is, instead of solving Schrödinger's (differential) equation one obtains the probability for a specific outcome by integrating an exponential function of the action over all of the worldlines from the source through the entire spatiotemporal configuration of experimental equipment to the detection event (Figure 9.2). Luckily, to appreciate our point you do not need to know specifically how that is done. All you need to know is that you must specify the entire spatiotemporal configuration of experimental equipment and the outcome(s) in order to compute the probability. So, in this 4D (integral) view one is asking about the distribution of entire spatiotemporal measurement configurations with specific outcomes in 4D spacetime (Figures 9.3 and 9.4).

Since we are computing the probability using the action for the worldlines of particle-like bodily objects, does that imply quanta have worldlines like bodily objects? No, because spatial location and momentum are simultaneously

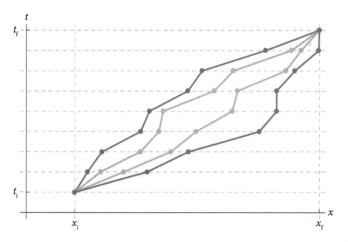

Fig. 9.2 In the path integral approach, one must compute the action over each and every possible particle path in spacetime (worldlines) from the source emission event to the detection event.

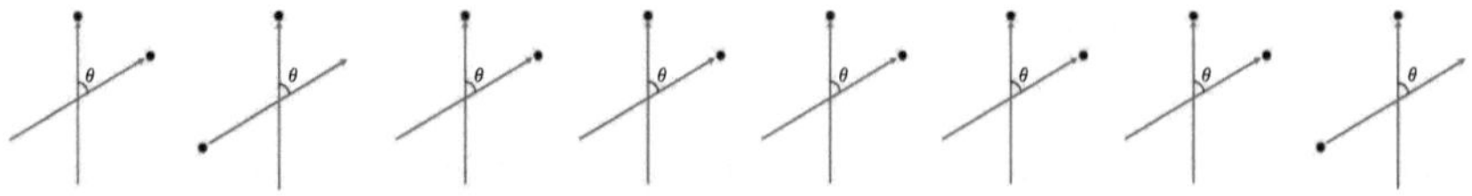

Fig. 9.3 An ensemble of eight spatiotemporal SG measurement trials of a Bell spin triplet state by Alice and Bob. The blue arrows depict Alice's and Bob's SG magnet orientations and the black dots represent their measurement outcomes for each trial, up (+1, located at arrow tip) or down (−1, located at bottom of arrow). So, each of the eight crossed blue arrows with black dots represents the triplet state source, Alice's and Bob's SG magnet orientations, and Alice's and Bob's outcomes for a specific trial of the experiment in spacetime. In these trials we are looking at Bob's (Alice's) outcomes corresponding to Alice's (Bob's) +1 outcomes when $\theta = 60°$. As we showed in Chapters 2 and 7 for triplet state measurements, the probability that Bob (Alice) obtains a +1 outcome when Alice (Bob) obtains a +1 outcome and the angle between their SG magnets is $\theta = 60°$ is $\cos^2(\theta/2) = 0.75$. That is why we see that Alice and Bob both obtained +1 in six of the eight spatiotemporal configurations (trials) of this experiment. The order that these eight spatiotemporal configurations is presented in this figure has nothing to do with when or where they were conducted in the larger context of spacetime globally (Figure 9.4).

available along the worldline of a particle-like bodily object and that cannot be true for quanta. The only place quanta can be located in our spatiotemporally global depiction of worldlines/worldtubes for bodily objects exchanging quanta would be as 4D regions of action h in the worldtubes of the emitters and receivers (as we explained and qualified in Subsection 9.3.1). Nonetheless, there has to be some consequence for the emission and absorption of quanta on the geometry of spacetime according to general relativity, since in general relativity the geometry of spacetime is determined by its matter–energy content.

If you are interested in reading more about how our proposed ontology bears on matters of gravity, read Subchapter 9B. If you are interested in seeing how our account of quantum mechanics resolves the mysteries of the double-slit experiment and delayed choice, read Subchapter 9C. Finally, if you are interested in seeing how our account of quantum mechanics resolves the mysteries of Schrödinger's Cat and Wigner's Friend, read Subchapter 9D. After reading any of these subchapters of interest, you can continue to Section 9.4 in this main chapter.

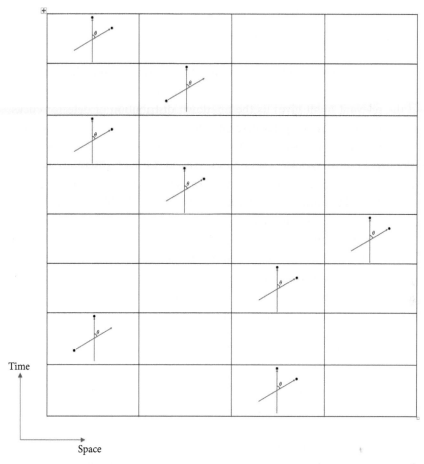

Fig. 9.4 A distribution of the ensemble of eight spatiotemporal SG measurement trials of a Bell spin triplet state by Alice and Bob (Figure 9.3) in the global spacetime context. As you can see, it makes no difference where along the Space axis or when along the Time axis the trial is placed.

9.4 An Ontology Based on the Poincare Symmetries

We have presented three experimental examples of Information Invariance & Continuity for the qubit, i.e., spin-$\frac{1}{2}$, where the mystery is 'average-only' projection; photon polarization, where the mystery is 'average-only' transmission; and the double-slit experiment (in Subchapter 9C), which has been called "The Quantum Experiment That Broke Reality" [26], where the mystery is wave–particle duality. All of these mysteries stem from the fact that measurements of h must produce the same value in all inertial reference frames, and h is an action.

For spin-$\frac{1}{2}$ we have angular momentum multiplied by angular displacement and spin angular momentum = $\hbar/2$. Angular momentum generates rotations and the relevant qubit gives us the up–down distribution of detector clicks (quantum events) in any particular direction of space as the SG magnets are rotated in continuous fashion. For photon polarization we have energy (E) multiplied by time and $E = hf$. Energy generates time translations and the relevant qubit gives us the distribution of detector clicks in time as the polarizer is rotated in continuous fashion. Finally, for the double-slit experiment we have momentum (p) multiplied by displacement and $p = h/\lambda$. Momentum generates spatial translations and the relevant qubit gives us the distribution of clicks in the interference pattern along the detector screen as the detector screen is spatially translated in continuous fashion.

Since the (finite-dimensional) Hilbert space formalism of quantum mechanics can be built from Information Invariance & Continuity for the qubit, quantum mechanics is a principle theory built upon the empirically discovered fact that h has the same value in all inertial reference frames. And, since this empirically discovered fact is easily justified by the relativity principle (NPRF), we see that quantum mechanics as a principle explanation is based on NPRF just like special relativity.

In special relativity we have the Michelson–Morley experiment with its mystery of length contraction. This mystery stems from the fact that measurements of c must produce the same value in all inertial reference frames to include those in relative motion and c is a velocity. Since velocity generates boosts, our two theories (Maxwell's equations and Planck's blackbody radiation equation) giving us two constants of Nature (c and h) that four experiments have shown to have the same value in all inertial reference frames (Michelson–Morley, Stern–Gerlach, photon polarization, and double-slit) reflect the Poincaré symmetries, the symmetry group of Minkowski spacetime (Figure 1.6).

So, again, NPRF + c constrains the spacetime configuration of worldtubes for bodily objects (with their unlimited simultaneous information) according to the invariant large-but-finite speed c of information transfer, while NPRF + h dictates the distribution of quanta (with their limited simultaneous information) among those bodily objects according to the invariant small-but-nonzero action h. It makes sense that something in the ontology other than bodily objects with their worldtubes defined by commuting momentum and position is responsible for the interaction of bodily objects. If bodily objects mediated the interaction between bodily objects, the mediating bodily objects would need interactions to establish their commuting momentum and position, etc., creating an infinite regress of ontological entities. That infinite regress is stopped before it can be started in this ontology because quanta are not bodily objects.

Concerning the distribution of quanta in the context of the worldtubes of bodily objects, we see that quantum mechanics can specify distributions in a direction of space (SG experiment) or along a spatial extent (multi-slit experiment) or in time (photon polarization experiment). And, we know that the commuting properties of any given bodily object are established by a multitude of quanta distributed in the context of other bodily objects. So, do we have any actual experiments showing how a bodily object is created by individual quanta? Besides the obvious fact that quanta produce worldlines in detectors (an individual click), we did get a worldline for a rubidium atom in an atomic trap via the exchange of 100 000 photonic quanta per second with photon detectors. But, far fewer quantum events are needed to statistically establish the worldline of a bodily object, e.g., in a particle physics experiment a worldline can be created by only two quantum events if the bodily object is not accelerating.

A particle physics "event" contains "approximately 100 000 individual measurements of either energy or spatial information" [18] (Figure 9.5). Individual detector clicks (called *hits in the tracking chamber*) are first localized spatially (called *preprocessing*), then associated with a particular worldline (called *pattern recognition*). The worldlines must then be parameterized to obtain dynamical characteristics (called *geometrical fitting*) [16, §§1.7.1, 1.7.2, and 1.7.3]. This is the process by which worldlines for particles in a high-energy physics experiment (bodily objects) are defined in the context of detectors (bodily objects). Since these particles are bodily objects, they are analyzed using relativistic classical mechanics (Figure 9.6).

Each worldline (or *track*) is deduced click (from a quantum event) by click (from another quantum event) in spatiotemporal succession, so we must obtain enough clicks for a worldline to find its acceleration from a radius of curvature (Figure 9.6). Obviously, a free particle would not accelerate and its worldline could be deduced statistically from only two spatiotemporally successive clicks (from different quantum events). For example, Nevill Mott showed that the path of an alpha particle emitted in a radioactive decay in a cloud chamber would be a straight line (again, statistically speaking) by using Schrödinger dynamics to compute the probability that two successive cloud

Fig. 9.5 "The event was recorded by ATLAS on 18-May-2012, 20:28:11 CEST in run number 203602 as event number 82614360." http://cds.cern.ch/record/1459495. Reproduced by permission of CERN.

chamber gas particles (hydrogen atoms for Mott) would be excited [24]. He found that [17]:

> the two hydrogen atoms have negligible probability to be both excited unless the atoms and the radioactive source lie on the same straight line. The result, ... implies that only straight tracks have non zero probability to be observed in the cloud chamber camera.

So, in the transition from quantum to bodily object, it is possible to statistically create 'simple' bodily objects defined contextually by tracks (worldlines) in particle detectors, and these simple bodily objects can be created with very few quantum events (as few as two for the alpha particle in Mott's analysis). The context is a particular source, externally applied magnetic and/or electric fields, and different types of detectors. Philip Goyal writes [20]:

> But, if we wish to entertain a kind of context-dependent persistence, we seem to have no choice but to give up the idea of 'particles' as basic. Seemingly the only recourse is to take the flashes themselves as basic, and thus to regard the

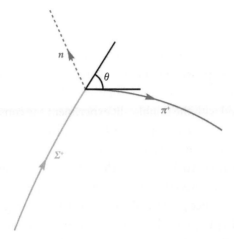

Fig. 9.6 "The particle decay $\Sigma^+ \rightarrow \pi^+ + n$ is observed in a bubble chamber. This figure represents the curved tracks of the particles Σ^+ and π^+ and the invisible track of the neutron in the presence of a uniform magnetic field of 1.15 T directed out of the page. The measured radii of curvature are 1.99 m for the Σ^+ particle and 0.580 m for the π^+ particle. From this information, we wish to determine the mass of the Σ^+ particle." This problem is from Raymond A. Serway and John W. Jewett, *Physics for Scientists and Engineers with Modern Physics*, 10th edition (2019).

notion of particle as a secondary notion—as a conceptual device for threading a sequence of flashes, a device that is only applicable in certain experimental contexts.

Quantum field theory provides a distribution of these simple bodily objects (particles) for each context. Indeed, Daniele Colosi and Carlo Rovelli make the point that the particles of high-energy physics are defined contextually by the experimental arrangement [10]. The key to finding the distribution and types of simple particles for a given context is to find the relevant action for quantum field theory which must be constructed from a Lagrangian that is ... Poincaré symmetric.

Notice that we have moved from the low-energy realm of quantum mechanics to the high-energy realm of particle physics, so our theory (quantum field theory) has Poincaré symmetry while quantum mechanics does not (it lacks Lorentz boost symmetry). Indeed, Schrödinger's equation is just the low-energy approximation to the Poincaré symmetric Klein–Gordon equation from quantum field theory [38, p. 172]. Recall, Mott used Schrödinger's equation for his low-energy analysis of excited hydrogen atoms to statistically produce the two-event worldline for the alpha particle. So, the ontology presented here based on the Poincaré symmetries is quite compatible with special relativity, quantum mechanics, and quantum field theory.

We conclude by emphasizing how this Poincaré symmetric ontology with AGCs represents a radical departure from typical constructive thinking. For example, confronted with the double-slit experiment the constructive thinker wants to know exactly what is passing through the slits. We end up with particle-like outcomes at the detector that are distributed in a wave-like interference pattern, so are particles passing through the slits or are waves? If it is particles, then they must pass through one slit or the other, so how do they end up creating an interference pattern? Perhaps you have waves guiding particles? If it is waves, then they impinge along the entirety of the detector screen, so how do they end up creating particle events? Perhaps it is spontaneous collapse? Or perhaps the waves do not collapse, but the universe splits into infinitely many worlds, each with a point-like outcome?

Here is a typical constructive response from Maudlin [19]:

> You would like a clear physical account of what's really going on ... The first thing you want to ask somebody when they're trying to explain the physics underlying this behavior is, "Do you have the first picture where in addition to the particle you have this wave-like thing (like Bohmian mechanics) or the second picture where you only have this wave-like thing and it collapses (spontaneous collapse models) or the third picture where you only have this wave-like thing and it doesn't collapse (like Many-Worlds)." Those are your three options.

It may be true that those are your only three options if you are looking for dynamical laws and/or mechanistic causal processes to solve the mystery of the double-slit experiment. But, as we explain in Subchapter 9C, there is an entirely different way to solve the mystery of the double-slit experiment.

That is, NPRF + h is an AGC that dictates we obtain quanta of momentum $p = h/\lambda$ in the inertial reference frames associated with the locations x along the interference pattern (as with bullets in Figure 9C.5), rather than a continuum of wave intensity (as with water waves in Figure 9C.6). However, the quantum measurement context is that for momentum p, so an interference pattern allowing us to compute λ for use in $p = h/\lambda$ is required. Therefore, the interference pattern per classical wave mechanics is what must obtain on average in this quantum-mechanical context.

In all three quantum experiments above (spin-$\frac{1}{2}$, photon polarization, double-slit experiment), NPRF + h demands that a classically continuous quantity (angular momentum, energy, momentum) be quantized, so that you have to take an average to recover the classical continuity. By accepting that principle explanation via AGCs can be fundamental to constructive explanation via causal mechanisms, these mysteries are accepted as unavoidable consequences of the relativity principle. This allows us to avoid the endless debate about which property of constructive explanation must be compromised, which is responsible for the current impasse in foundations of quantum mechanics. Salom writes [32]:

> We hope that the Nobel prizes awarded last year [2022] can help raise awareness of the scientific and philosophical revolution that happened almost half a century ago (with the first Bell tests) and that a clockwork mechanism picture of reality will soon be widely recognized for what it is—a "flat Earth" of philosophy.

9.5 God's Thoughts in Creating the World

At the beginning of this chapter, we told you we would speculate freely to answer Zeilinger's question, "What's really going on?", Becker's question, "What is real?", and Einstein's question, "What were God's thoughts when He created the world?" In this final section, we will summarize that speculation to highlight the profound consequences for your everyday thinking.

Einstein is not alone in using the God metaphor when it comes to the question of creation and the nature of reality. What follows is Barry Loewer's description of a widely held constructive answer to these questions [22]:

> Ok; so what did God have to do to create our universe? One answer goes like this. What God did first ... was to create Space-Time ... Second, God decided on the types of fundamental material ontology—particles, fields, strings, wave functions and their fundamental properties [like] charge, charm, and so forth. Third, God created the laws that govern the distribution of the fundamental ontology and quantities in space and time (or the chances of such distributions if God was so perverse as to play dice with the universe by decreeing indeterministic laws). Fourth, God selected a particular distribution of the fundamental ontology and properties to be the initial condition of the universe. Of course, God had to make sure that Space-Time, the ontology,

the laws, and the initial condition were all [consistent with each other]. This is a very hard problem for physicists—think of making relativity and quantum mechanics compatible—but we can suppose that God, being omniscient, had no trouble with it. And last, but not least, God ... started Time going—or flowing.

If you are interested in a more detailed account of how physics actually fills in the preceding constructive picture of reality, you can read Sean Carroll's recent book, *The Biggest Ideas in the Universe: Space, Time, and Motion* [7].

As you can see, right in the middle of Loewer's story is the acknowledgment of "a very hard problem for physicists": how to make the dynamical view of reality self-consistent. There are many such inconsistencies in the dynamical view of reality [34], but the one we have focused on in this book was mentioned specifically by Loewer, i.e., "think of making relativity and quantum mechanics compatible." If Loewer's constructive account represents God's thoughts in creating the world, then we would agree with Maudlin that these are "the ways of a Deity who is, if not evil, at least extremely mischievous."

But, what we have shown you in this book is an entirely different conception of reality according to which God was not malicious in creating the world, but benevolent. In our principle, constraint-based conception of the world, God's design plan was very simple and impartial—the laws of physics and their constants will be the same in every inertial reference frame. So, the principle constraint ("God's thought") is the relativity principle, or no preferred reference frame (NPRF), and it constrains the distribution of events in spacetime as a whole, whether they are quantum events (constrained by NPRF + h) or classical events (constrained by NPRF + c).

NPRF is God's AGC on how our individual, dynamical ((3+1)D) subjective spacetime models of reality can be reconciled in a single intersubjectively-agreed-upon, all at once (4D) objective spacetime model of reality per Einstein's "game" of physics. That is, the AGC on events in the 4D all-at-once objective spacetime model of reality in no way negates the fact that our subjective experience truly reflects a dynamical reality [33, 34]. In other words, nothing we have said entails the B-theory of time. Any particular individual, (3+1)D dynamical story (for a conscious or inanimate data collection device) can be read off from the 4D all-at-once objective spacetime model of reality.

If we rather try to create a dynamical/causal objective spacetime model of reality, our model will necessarily violate locality, statistical independence, intersubjective agreement, and/or unique experimental outcomes as necessary to account for the violation of Bell's inequality. In other words, some cherished characteristic of constructive explanation must be sacrificed. So, an obvious question is, "Why would anyone force the objective spacetime model of reality to be dynamical?" [Pun intended.]

In short, special relativity and quantum mechanics have shown us that, unlike Newtonian mechanics where objective reality (reality) aligns perfectly with our dynamical/causal subjective experience, reality differs profoundly from our subjective experience in two respects:

(i) The best way to unify the empirical data from our data collection devices in different inertial reference frames is not via dynamical laws or mechanistic causal processes as in Newtonian mechanics. Rather, reality is governed most fundamentally by a principle of impartiality, i.e., the relativity principle.

(ii) As a result of this principle, reality is best understood in all-at-once rather than time-evolved fashion.

Accordingly, Frank Wilczek issued the following challenge in 2016 [37, p. 37]:

A recurring theme in natural philosophy is the tension between the God's-eye view of reality comprehended as a whole and the ant's-eye view of human consciousness, which senses a succession of events in time. Since the days of Isaac Newton, the ant's-eye view has dominated fundamental physics. We divide our description of the world into dynamical laws that, paradoxically, exist outside of time according to some, and initial conditions on which those laws act. The dynamical laws do not determine which initial conditions describe reality. That division has been enormously useful and successful pragmatically, but it leaves us far short of a full scientific account of the world as we know it. The account it gives—things are what they are because they were what they were—raises the question, Why were things that way and not any other?

The God's-eye view seems, in the light of relativity theory, to be far more natural. Relativity teaches us to consider spacetime as an organic whole whose different aspects are related by symmetries that are awkward to express if we insist on carving experience into time slices.

To me, ascending from the ant's-eye view to the God's-eye view of physical reality is the most profound challenge for fundamental physics in the next 100 years. [Emphasis added.]

This is a world of relative simultaneity and 'average-only' conservation. Without taking this impartial principle of creation into account, an entirely unnecessary tension between special relativity and quantum mechanics results. It is that simple. It turns out that Einstein was spot on when he said [24]:

Subtle is the Lord, but malicious He is not.

If only Einstein had again abandoned his "constructive efforts" and rejected Reichenbach's Principle in favor of his beloved relativity principle, he could have derived the Hilbert space of quantum mechanics and avoided the mystery of entanglement altogether. Einstein's entanglement is certainly steeped in irony.

In conclusion, let us remind you of Wheeler's prediction quoted at the outset of Chapter 1 [36]:

The necessity of the quantum in the construction of existence: out of what deeper requirement does it arise? Behind it all is surely an idea so simple, so beautiful, so compelling that when—in a decade, a century, or a millennium—we grasp it, we will all say to each other, how could it have been otherwise? How could we have been so stupid for so long?

Amen to that, John.

References

[1] E. Adlam, *Two Roads to Retrocausality*, 2022. Preprint. https://arxiv.org/abs/2201.12934.

[2] E. Adlam, 2023. Personal correspondence.

[3] M. Araújo, F. Costa, and C. Brukner, *Computational Advantage from Quantum-Controlled Ordering of Gates*, Physical Review Letters, 113 (2014), p. 250402.

[4] V. Baumann, M. Krumm, P. A. Guérin, and C. Brukner, *Noncausal Page–Wootters circuits*, Physical Review Research, 4 (2022), p. 013180.

[5] A. Becker, *What is Real? The Unfinished Quest for the Meaning of Quantum Physics*, Basic Books, New York, 2018.

[6] R. Bishop, M. Silberstein, and M. Pexton, *Emergence in Context*, Oxford University Press, Oxford, 2022.

[7] S. Carroll, *The Biggest Ideas in the Universe: Space, Time, and Motion*, Dutton, New York, 2022.

[8] G. Chiribella, G. M. D'Ariano, P. Perinotti, and B. Valiron, *Quantum computations without definite causal structure*, Physical Review A, 88 (2013), p. 022318.

[9] J. Chu, *Scientists set traps for atoms with single-particle precision*, MIT News, 3 November (2016). https://news.mit.edu/2016/scientists-set-traps-atoms-single-particle-precision-1103.

[10] D. Colosi and C. Rovelli, *What is a particle?*, Classical and Quantum Gravity, 26 (2009), p. 025002.

[11] E. Crull, *Physics Scratches a Philosopher's Itch*, Physics, 15 (2022), p. 73.

[12] C. Day, *Quantum information is exciting and important*, Physics Today, 74 (2021), p. 8.

[13] A. Einstein, *Physics and Reality*, Journal of the Franklin Institute, 221 (1936), pp. 349–382.

[14] Energy.gov, *White House Office of Technology Policy, National Science Foundation and Department of Energy Announce Over $1 Billion in Awards for Artificial Intelligence and Quantum Information Science Research Institutes*, 2020. https://www.energy.gov/articles/white-house-office-technology-policy-national-science-foundation-and-department-energy.

[15] M. Esfeld and N. Gisin, *The GRW flash theory: A relativistic quantum ontology of matter in space-time?*, 2013. Preprint. https://arxiv.org/abs/1310.5308.

[16] R. Fernow, *Introduction to Experimental Particle Physics*, Cambridge University Press, Cambridge, 1986.

[17] R. Figari, *Particle tracks in a cloud chamber: the Mott's conjecture (1929)*, Proceedings of the 36th National Congress of the Italian Society for the History of Physics and Astronomy, (2017), pp. 331–337.

[18] H. J. Frisch, *Pattern recognition at the Fermilab collider and Superconducting Supercollider*, Proceedings of the National Academy of Sciences, 90 (1993), pp. 9754–9757.

[19] M. Geleta, *Tim Maudlin: Philosophy of Science and Quantum Physics*, 2023. https://www.youtube.com/watch?v=40uw7jNo4_8.

[20] P. Goyal, *Persistence and Reidentification in Systems of Identical Quantum Particles: Towards a Post-Atomistic Conception of Matter*, 2023. Preprint. http://philsci-archive.pitt.edu/21849/.

[21] P. Lewis and M. Silberstein, *Conservation laws and quantum chance* (2024). https://philsci-archive.pitt.edu/.

[22] B. Loewer, *Time and Law*, 2004. https://sites.rutgers.edu/barry-loewer/wp-content/uploads/sites/195/2019/06/Loewer-Time-and-Law.pdf.

[23] T. Maudlin, *Quantum Non-Locality and Relativity*, Wiley-Blackwell, Oxford, 2011.

[24] P. Morrison, *Review of* Subtle Is the Lord...: The Science and the Life of Albert Einstein, Scientific American, 248 (1983), pp. 30–37.

[25] N. F. Mott, *The wave mechanics of alpha-ray tracks*, Proceedings of the Royal Society of London A: Mathematical, Physical and Engineering Sciences, 126 (1929), pp. 79–84.

[26] PBS Space Time, *The Quantum Experiment That Broke Reality*, 2016. https://www.youtube.com/watch?v=p-MNSLsjjdo.

[27] J. Preskill, *Entanglement Explained!*, 2020. https://online.kitp.ucsb.edu/online/qgravityt-c20/preskill/.

[28] C. Rovelli, *Reality Is Not What It Seems: The Journey to Quantum Gravity*, Riverhead Books, New York, 2017.

[29] C. Rovelli, *Helgoland: Making Sense of the Quantum Revolution*, Riverhead Books, New York NY, 2021.

[30] G. Rubino, L. A. Rozema, A. Feix, M. Araújo, J. M. Zeuner, L. M. Procopio, C. Brukner, and P. Walther, *Experimental verification of an indefinite causal order*, Science Advances, 3 (2017), p. e1602589.

[31] S. Saito, *Spin of Photons: Nature of Polarisation*, 2023. Preprint. https://arxiv.org/abs/2303.17112v1.

[32] I. Salom, *2022 Nobel Prize in Physics and the End of Mechanistic Materialism*, Phlogiston, 31 (2023). In Press. https://arxiv.org/abs/2308.12297.

[33] M. Silberstein, *Commentary: Physical time within human time, Frontier of Psychology*, 14 (2023). https://doi.org/10.3389/fpsyg.2023.1096280.

[34] M. Silberstein, W. M. Stuckey, and T. McDevitt, *Beyond the Dynamical Universe: Unifying Block Universe Physics and Time as Experienced*, Oxford University Press, Oxford, 2018.

[35] S. Weinberg, *Reductionism Redux*, The New York Review of Books, 5 October (1995). https://www.nybooks.com/articles/1995/10/05/reductionism-redux/.

[36] J. A. Wheeler, *How Come the Quantum?*, New Techniques and Ideas in Quantum Measurement Theory, 480 (1986), pp. 304–316.

[37] F. Wilczek, *Physics in 100 Years*, Physics Today, 69 (2016), pp. 32–39.

[38] A. Zee, *Quantum Field Theory in a Nutshell*, Princeton University Press, Princeton, 2003.

[39] B. Zwiebach, *Photons and the loss of determinism* (2017). https://www.youtube.com/watch?v=8OsUQ1yXCcI.

9A

Interpreting Quantum Mechanics in a Realist Fashion

What exactly makes an interpretation of quantum mechanics a realist one is up for debate, but at the very least it is generally believed that realist interpretations cannot be purely epistemic. For example, according to QBism, quantum probabilities are not about objective reality, rather they are about updating the belief states of single agents. Richard Healey's pragmatist account of quantum mechanics and Rovelli's relational quantum mechanics both hold that the quantum state is not a description of the physical world, but only exists to generate quantum probabilities for the purposes of making predictions about outcomes. Unlike QBism, both pragmatist accounts and relational accounts of quantum mechanics are relative-state theories, i.e., in the pragmatist account a quantum state is relative only to the perspective of an actual or potential agent, and in relational quantum mechanics values are information a physical system has about another system, as with a third physical system observing two other entangled systems.

In the original version of relational quantum mechanics, all information is purely relational and this need have nothing to do with 'agents' even in the neutral sense of agent in which a non-conscious device might be observing and measuring outcomes in quantum experiments. However, Healey is clear that a quantum state can only be ascribed to an agent in the context of an experimental setup that defines the perspective of that agent; in this respect Healey's pragmatist quantum mechanics is a kind of halfway house between QBism and relational quantum mechanics. In spite of the agent-centric talk in QBism and the pragmatist account, these accounts do not require conscious agents, i.e., an epistemic agent could be a non-conscious machine of some sort.

What makes all three of these accounts epistemic is that they all hold that the quantum state as given by the wavefunction is not a description of the physical world, but only exists to generate quantum probabilities. That is, in addition to being explicitly psi-epistemic, none of these accounts provides an ontology or what Bell called *beables* (localized fundamental physical entities

Einstein's Entanglement. W. M. Stuckey, Michael Silberstein, and Timothy McDevitt, Oxford University Press.
© Oxford University Press (2024). DOI: 10.1093/9780198919698.003.0011

such as particles whose properties are faithfully measured in experiments) that are allegedly hiding behind the veil of the "observables" (such as measurement outcomes, interference patterns, etc.) and explain the observable phenomenology in question. The beables, such as particles, fields, or waves of some sort, are supposed to tell us exactly what happens, say, between the initiation and termination of some quantum experiment, such as a Bell-type or double-slit-type experiment.

In some constructive interpretations of quantum mechanics the beables represent what Allori [2] and others call a *primitive ontology*, i.e., one that cannot be inferred from the formalism of relativistic or non-relativistic quantum mechanics. So, a primitive ontology is more fundamental than even textbook quantum particles, fields, or waves. The motivation for a primitive ontology is the same as for positing local beables [4, Chapter 7] more generally. The goal is to provide a constructive ontology in spacetime that can account for definite measurement outcomes, EPR–Bell correlations, and the existence of classical, localized objects such as measuring devices.

It is sometimes further claimed that beables must have some metaphysical autonomy/independence (their existence requires no interactions with anything else) and some intrinsic properties that are faithfully and objectively represented by measurements (such properties and their values do not only exist relative to a measuring device in a certain setting). If that is so, then earlier versions of relational quantum mechanics fail to be a realist theory. As Federico Laudisa and Rovelli put it [6]:

> For RQM (relational quantum mechanics), the lesson of quantum theory is that the description of the way distinct physical systems affect each other when they interact (and not the way physical systems 'are') exhausts all that can be said about the physical world. The physical world must be described as a net of interacting components, where there is no meaning to 'the state of an isolated system,' or the value of the variables of an isolated system. The state of a physical system is the net of the relations it entertains with the surrounding systems. The physical structure of the world is identified as this net of relationships.

So, if relational quantum mechanics is true, there are no such things as beables so defined. The same is true for our view as well.

The ontology in Rovelli's relational quantum mechanics is 'sparse' in the sense that properties are instantiated at spacetime points. But it is also relational—the properties so instantiated are relations between systems that interact at that point. And the properties need not be position properties—the property that is instantiated depends on the nature of the interaction. The wavefunction is again law-like—given a relational property at one location, it tells you the probabilities of relational properties at other locations. The relationality solves the problem with entangled particles—there is Alice's spin relative to Alice, and Bob's spin relative to Bob, but no requirement that there be any correlations between them, so no need for nonlocality as in "spooky action at a distance." But, as alluded to above, it comes with a big problem. The relationality means that Alice can never know anything about Bob's results, in principle, or her own results five minutes ago.

Rovelli's relational quantum mechanics entails that there is no sense asking whether Alice and Bob saw the same result—there is just the result relative to Alice, and the result relative to Bob. Relational quantum mechanics implies that the claim "Alice and Bob got the same result" means "If Alice asks Bob for his result, his utterance relative to Alice's agrees with the result relative to a third observer Ted, who is observing both Alice and Bob." Here is how Rovelli puts it [8]: "The joint properties of two objects exist only in relation to a third ... the [quantum] correlation manifests itself when the two correlated objects both interact with this third object [measuring device]."

Second, Rovelli's relative-state relationalism is robust against changes in perspective—unlike frame-dependent properties in special relativity, any inconsistency is equally there in the 4D perspective, the all-at-once perspective. Third, as discussed, relational quantum mechanics says all properties are relative to the "quantum state" of interacting "observers." Again, as explained in Subchapter 9D, this means relational quantum mechanics entails that there can be contradictions between what observers report about the outcomes of numerically identical quantum experiments, which cannot be reconciled or explained away. This is the so-called Wigner's Friend problem or failure of intersubjective agreement mentioned in the main chapter. We discuss this problem at length in Subchapter 9D with a suggested fix offered by Adlam and Rovelli.

As we show in Subchapter 9D, our view with its quantum–classical contextuality does not have this problem. This is because, as discussed above, our contextual emergence essentially includes the classical measuring device. We will show that the Wigner's Friend problem only arises when one illegitimately tries to treat classical things (such as observers measuring observers measuring stuff and their records) as quantum mechanical in the vein of relational states or subjective collapse.

As noted, another criterion for a theory to be a realist one is that it does not entail a violation of intersubjective agreement, that is, the existence of objective, observer-independent reality. Some accounts, such as the afore-mentioned original version of relational quantum mechanics, are labeled subjectivist because they allegedly entail that, at least in certain situations (e.g., Wigner's Friend), different observers can consistently give different accounts of the same set of events such as the outcomes of measurements. For example, in a particular Schrödinger's Cat setup, even without invoking the branching structure of the Many-Worlds interpretation, observer X can report seeing a live cat and observer Y can report seeing a dead cat, allegedly the very same cat, and both can be correct without contradiction [3, pp. 116–117]. That is, such subjectivist accounts allegedly violate what is sometimes called The Absoluteness of Observed Events [9].

This consequence of the original version of relational quantum mechanics is thought to derive from its relationality. That is why Adlam and Rovelli [1] recently modified relational quantum mechanics by adding a new postulate designed to block any failure of intersubjective agreement in cases like Wigner's Friend. In Subchapter 9C on delayed choice and Subchapter 9D on Schrödinger's Cat and Wigner's Friend, we will show how intersubjective agreement is maintained by our 4D adynamical global constraints ranging over quantum–classical contextuality. What will become clear is that contextuality of the sort suggested by the Kochen–Specker theorem does not entail the kind of relationality of facts to "observers" that got relational quantum mechanics in trouble.

In this chapter it will become clear that in our realist all-at-once approach there is no need to posit any primitive ontology. As Healey [5] notes, for example, quantum mechanics does not have an ontology of its own, i.e., it just piggybacks on whatever classical theory you are quantizing. So, standard non-relativistic quantum mechanics relativizes Newtonian particle mechanics: the ontology is point-like. Relativistic quantum field theory relativizes a field theory and makes it discrete, so the ontology is field values at spacetime points. Consequently, no new ontology is needed unless one is a wavefunction fundamentalist or has some other motivation for believing there must be underlying primitive beables not presently in physics. In our all-at-once view, there is no need to be a wavefunction realist, nor is there any need for a primitive ontology. And again, there is no reason that particles and fields must be beables in the sense of violating quantum–classical contextuality.

As will be clear in what follows, our ontology is straightforwardly about bodily (classical) objects, such as sources and measuring devices, that interact via the exchange of quanta, which one is free to characterize as quantum

particles or fields without worldlines. As discussed, given our multiscale contextual emergence it would be a mistake to say that classical objects are
literally 'composed' of quanta in some synchronic, Lego-like fashion. But
again, atoms are not literally composed of subatomic particles regardless of
one's ontology. And, it is an empirical fact that quantum behavior differs
from classical behavior by virtue of the presence or absence of interactions
with the external environment, regardless of size. For example, large individual atoms have been made to exhibit entanglement, as have clouds of cold
gases made of thousands of atoms and oscillators containing roughly 10^{10}
atoms [7].

All of this demonstrates that neither quantumness nor classicality need be
absolute or be primarily about scale, it is about context. The point is that
classical bodily objects and their exchanges of quanta are all that is needed.
Furthermore, in this chapter, by adding the multi-slit experiment to the other
two experimental instantiations of the qubit, i.e., the SG experiment and photon polarization, we show that the Poincaré symmetries govern the behavior
of both quanta and the classical.

In order to make sure the reader appreciates our ontological picture of reality and why it is not instrumentalism, it is important to contextualize what is
to follow. To reiterate, in our view the following are true:

(i)   There cannot be quanta (defined as noncommutative, no worldlines)
      without classical objects (defined as commutative, with worldlines).
(ii) There cannot be classical objects without quanta.
(iii) There are no possible worlds with just one quantum or just one
      classical object.
(iv) The 4D distribution of classical objects and their direct (no world-
      lines) 'exchanges' via quanta is given by our adynamical or acausal
      global constraints (AGCs) as discussed.

Again, keep in mind these 'exchanges' (of physical information if you like)
are not causal or dynamical in the sense there are no worldlines and in the
sense that the 4D distribution of quanta represents a 4D pattern as given by
the AGCs. As stipulated, the exchanges of quanta are exactly the properties
given to us by physics such as energy and momentum, etc. There is no primitive ontology. Again, all of this characterizes what we call quantum–classical
contextuality.

If we have failed to make things clear thus far, you may ask, how can we
just 'give ourselves' classical objects if they are made of quanta and thus not
fundamental? Is it not the case that classical objects are made of atoms which

in turn are composed of quanta, i.e., subatomic particles? As already noted, however, for a few reasons, we do not think this is the ultimate way to conceive of reality. First, as we will demonstrate, whether an atom behaves like a quantum or behaves like something classical is a function of 4D quantum–classical contextuality. Two examples of such contextuality are given by the double-slit experiment in Subchapter 9C and "a source of alpha particles of momentum p" in the main chapter. Indeed, whether or not some device behaves as a source or detector is a matter of 4D contextuality, but this does not make such things anthropocentric. There is no odious sense in which measurement is fundamental in our view.

Second, invoking classical objects as in some sense basic is not cheating, because most of the assumptions that would lead to this worry do not apply in our view. For example, quantum mechanics as viewed constructively is not fundamental, the fundamental explanation for classical objects is not constructive or dynamical, and wavefunction realism is false in our view. Indeed, for us the quantum is just NPRF + h, which dictates the kinematic structure of quantum mechanics (Hilbert space viewed as an adynamical constraint) exactly like NPRF + c dictates the kinematic structure of special relativity (Minkowski spacetime). And as we showed, the Hilbert space kinematic structure gets you quantum entanglement. This plus the Lagrangian account of quantum field theory (exchanges of quanta) is all there is to quantum mechanics. None of our constraints, NPRF + h, Lagrangians, etc., are inherently quantum or classical. So, quantum mechanics is not fundamental to the classical in the way many people assume that it must be, and there is no mystery about how the 'sober' classical with its definite values 'comes into being' from the ghostly potentia of quantum processes.

What is fundamental is 4D quantum–classical contextuality as given by all-at-once AGCs. If one fully embraces a 4D all-at-once conception of reality wherein classical objects do not dynamically 'emerge' from underlying quantum processes, there is no reason to view quantum–classical contextuality as a cheat. We just need to explain 4D distributions of mass–energy via AGCs. Again, for us, the quantum is just the kinematics of Hilbert space and those exchanges of quanta conceived in a Lagrangian fashion. We understand the reader may not appreciate all this now but it will all be clear by the end of this chapter. The key concept here is 4D constraints and not causal processes or dynamical laws. In our previous book [10] we discuss how this way of thinking can address many other problems in physics such as the origin of the Big Bang (Subcapter 9B) and the so-called grandfather paradox.

In the remainder of this chapter, we will spell all this out in more detail and show how our view enables a realist account of quantum mechanics that explains many puzzling quantum experiments involving quantum entanglement. We also explain how this new conception of the quantum bears on matters of gravity (Subchapter 9B).

References

[1] E. Adlam and C. Rovelli, *Information is Physical: Cross-Perspective Links in Relational Quantum Mechanics*, 2022. https://arxiv.org/abs/2203.13342.

[2] V. Allori, *Primitive Ontology in a Nutshell*, International Journal of Quantum Foundations, 1 (2015), pp. 107–122.

[3] J. Baggott, *Quantum Reality: The Quest for the Real Meaning of Quantum Mechanics—A Game of Theories*, Oxford University Press, Oxford, 2020.

[4] J. Bell, *Speakable and Unspeakable in Quantum Mechanics*, Cambridge University Press, Cambridge, 2nd ed., 2004.

[5] R. Healey, *The Quantum Revolution in Philosophy*, Oxford University Press, Oxford, 2017.

[6] F. Laudisa and C. Rovelli, *Relational quantum mechanics*, in Stanford Encyclopedia of Philosophy, E. N. Zalta, ed., Stanford University, 2019.

[7] S. Ornes, *Quantum effects enter the macroworld*, Proceedings of the National Academy of Sciences, 116 (2019), pp. 22413–22417.

[8] C. Rovelli, *Helgoland: Making Sense of the Quantum Revolution*, Riverhead Books, New York, 2021.

[9] M. Silberstein and W. M. Stuckey, *The Completeness of Quantum Mechanics and the Determinateness and Consistency of Intersubjective Experience: Wigner's Friend and Delayed Choice*, in Consciousness and Quantum Mechanics, S. Gao, ed., Oxford University Press, Oxford, 2022, pp. 198–259.

[10] M. Silberstein, W. M. Stuckey, and T. McDevitt, *Beyond the Dynamical Universe: Unifying Block Universe Physics and Time as Experienced*, Oxford University Press, Oxford, 2018.

9B

Gravity Matters

We need to figure out how to combine quantum mechanics and its relativistic (high-energy) version called quantum field theory with general relativity in order to understand how quanta couple to spacetime. This is what we meant when we said physics is not finished. The combination of quantum physics with general relativity is called *quantum gravity* and has been called "the hardest problem in physics" [13]. To appreciate our approach to quantum gravity that is consistent with the accounts of quantum mechanics and special relativity herein, we need to explain general relativity in a bit more detail.

General Relativity

Einstein's equations of general relativity are solved everywhere in spacetime for the metric $g_{\alpha\beta}$ and stress–energy tensor $T_{\alpha\beta}$,

$$G_{\alpha\beta} = \frac{8\pi G}{c^4} T_{\alpha\beta},$$

where G is Newton's gravitational constant (recall from basic physics) and $G_{\alpha\beta}$ is called the Einstein tensor, which is a very complicated function of $g_{\alpha\beta}$ to include its first and second derivatives in the four coordinates of spacetime. In fact, the left-hand side of Einstein's equations has thousands of $g_{\alpha\beta}$ terms when written in its most general form.

The metric tells you the geometry of spacetime, because it tells you how to find spatiotemporal distance in spacetime, while the stress–energy tensor tells you how matter, energy, and momentum are distributed in spacetime. Since energy and momentum require a knowledge of velocity and acceleration, which in turn require a knowledge of spatial distances and temporal durations, you cannot know the stress–energy tensor unless you know the metric. But, Einstein's equations say that you cannot know the metric unless you know the stress–energy tensor. In other words, Einstein's equations of general relativity tell you whether a metric and stress–energy tensor pair is possible.

Since Einstein's equations are so complex, exact solutions are obtained only for highly idealized matter–energy distributions. For example, the first exact

Einstein's Entanglement. W. M. Stuckey, Michael Silberstein, and Timothy McDevitt, Oxford University Press.
© Oxford University Press (2024). DOI: 10.1093/9780198919698.003.0012

solution was obtained by Karl Schwarzschild in 1916 (the year after Einstein introduced general relativity) and it gives $g_{\alpha\beta}$ outside a perfectly spherical, nonrotating mass M. Since his solution is for the region outside of M where there is no matter–energy, the Schwarzschild solution is called a *vacuum solution*.

According to the Schwarzschild solution, the spacetime curvature around Earth means that the clocks on Global Positioning System (GPS) satellites in orbit about Earth run a bit faster than those on the surface of Earth. The difference is very small, i.e., the satellite clocks are only 38.4 μs ahead of those on Earth after one day, but since locations are triangulated using light signals, which travel very fast, that small difference in clock time means our GPS locations would drift off by 11.5 km each day if not corrected [2].

Obviously, with such small changes to the spacetime geometry for a mass as large as Earth, we do not expect to find measurable deviations of spacetime curvature due to the exchange of low-energy quanta per quantum mechanics. If that did happen, we would be seeing violations of the predictions of quantum mechanics, since quantum mechanics does not take into account spacetime curvature. That is why we do not need a theory of quantum gravity for most phenomena.

Our Approach to Quantum Gravity

There is a straightforward approach to quantum gravity based on the co-fundamentality of the classical and the quantum in our quantum–classical contextuality. For example, to compute the probability of a particular outcome in a particular 4D experimental context using the path integral formalism (Subsection 9.3.2), one uses the action for each and every worldline from the (fixed) source emission event to the (fixed) quantum detection event. General relativity has an action formulation, so one could simply replace the action for the worldline in flat spacetime of Newtonian mechanics or special relativity with the action for the worldline in the curved spacetime of general relativity. Again, that difference will not be measurable for most situations (or we would have noticed deviations from the predictions of quantum mechanics already), but one can use it for any situation to include those where a measurable effect might be expected.

Of course, this is only dealing with quanta in the classical curved spacetime of general relativity, which is not what most people mean by quantum gravity. Following the analogy with the quantization of continuous classical fields and bodily objects as in the three experiments of the book, one might ask whether or not the continuous classical spacetime of general relativity is also quantized in certain contexts. Along those lines, let's imagine a hypothetical gravitational counterpart to the double-slit experiment for quanta of electromagnetic radiation and matter.

The Laser Interferometer Gravitational-Wave Observatory (LIGO) has shown that classical gravity exhibits wave-like behavior per general relativity [18], so this classical gravitational radiation should produce an interference pattern when encountering 'double slits.' In this idealized sense, some configuration of cosmic objects between Earth and the distant source of gravity waves would act as the 'double slits' and we would observe a variation in the amplitude of LIGO arm oscillations as the LIGO location moves through space (with Earth) along the interference pattern. Again, this is purely (ridiculously) hypothetical because gravity waves are so weak that LIGO has only measured bursts of radiation from cataclysmic events like the merger of two black holes [5], so we have no idea where to find a continuous source of gravity waves as necessary to observe gravitational wave interference patterns as described here. But, our ignorance notwithstanding, let's imagine that such an experiment could be carried out.

What we expect from quantum gravity in such an experiment would be discrete pulses in the LIGO arms associated with quanta of gravitational energy due to very small changes of matter configurations in the source. In other words, there must exist minimum changes to spatiotemporal relations (see below) between bodily objects in some contexts that lead to minimum amounts of gravitational potential energy just like photons for the electromagnetic field. In the weak field, linearized approximation of general relativity, these quanta of gravitational energy (called *gravitons*) have energy $E = hf$, just like any other quanta [21]. Accordingly, we expect that when such quanta of gravitational energy 'pass through' the cosmic 'double slits' they would not spread out in the interference pattern per classical wave theory, but would be localized spatiotemporally with a distribution that reproduces the classical interference pattern, just like quanta of electromagnetic radiation or matter in the double-slit experiment (see Subchapter 9C).

However, since the energy of the gravitational radiation being detected by LIGO is extremely small, a given graviton in those waves would have miniscule energy. For example, the frequency of the gravitational waves detected by LIGO is about 100 Hz [10, 17] while the frequency of red light is 4.3×10^{14} Hz. That means a photon of red light has about a trillion times more energy than a graviton in a LIGO gravitational wave. So, it is not likely that individual gravitons will be detected in the near future.

However, there might be observable consequences of gravitational quantum phenomena at the classical level, analogous to blackbody radiation for electromagnetic phenomena. Alexandre Deur proposed exactly that idea in 2009 when he used graviton–graviton scattering (quantum gravity phenomenon) to explain galactic rotation curve profiles (classical phenomenon) without a need for dark matter, i.e., matter that interacts via gravity but not electromagnetism [6]. He has since extended this approach to account for dark matter profiles in elliptical galaxies, galactic clusters, and the Bullet Cluster [7], the phenomenon of dark energy without a cosmological constant [8] (see below), and the discrepancy in the value of the Hubble constant in different measurements thereof [19].

Dealing with Singularities

Of course, there is another expectation of quantum gravity: that quantum gravity will replace the singularities in general relativity solutions with non-singular, quantum alternatives. Singularities are places in spacetime where the general relativity solution gives infinities (usually, but see below), and for those who believe the quantum is fundamental to the classical, e.g., bodily objects are composed of quanta, such singularities in general relativity solutions are expected to be replaced by the 'real,' well-behaved (quantum) solution in the singular region. We have two responses to that expectation.

First, consider the initial singularity of general relativity's Big Bang cosmology solution for the flat, dust-filled universe. The simplifying assumption for the FLRW metric for Big Bang cosmology (so named for its cofounders, Friedmann, Lemaître, Robertson, and Walker) is that spacetime can be foliated into spatial hypersurfaces of homogeneity (the same at every location) and isotropy (the same in every direction) leading to a general metric form with just one unknown function of one variable, $a(t)$. This scaling factor $a(t)$ gives the distance between fixed spatial coordinate positions at time t corresponding to spatial locations that are at rest with respect to the homogeneous, isotropic, space-filling dust.

Einstein's equations tell us that the spatial hypersurfaces can be spherical, flat, or hyperbolic, and the space-filling matter–energy then expands ($a(t)$ increasing with increasing time t, so density is decreasing) or contracts ($a(t)$ decreasing with increasing time t, so density is increasing). Einstein's equations give two differential equations for $a(t)$ in the flat, dust-filled model (the Einstein–de Sitter model). The first equation is

$$\frac{3\dot{a}^2}{a^2} = \frac{8\pi G\rho}{c^2}, \tag{9B.1}$$

where $\cdot = d/dt$ and $\rho(t)$ is the dust density at time t. The second equation is

$$2\ddot{a}a + \dot{a}^2 = 0. \tag{9B.2}$$

Equation 9B.2 can be written

$$\frac{d}{dt}(\dot{a}^2 a) = 0, \tag{9B.3}$$

which means $\dot{a}^2 a$ is a constant. Multiplying Eq. 9B.1 by a^3 then tells us that ρa^3 is a constant, which makes sense—the total mass in an expanding coordinate volume of dust-filled space is fixed. Continuing, we separate and integrate Eq. 9B.3 to give

$$\int_{a(0)}^{a(t)} \sqrt{a}\,da = At, \tag{9B.4}$$

where A is an arbitrary constant. The second arbitrary constant for the general solution of our second-order differential equation Eq. 9B.2 is of course $a(0)$.

Typically, $a(0)$ is chosen to be zero and $a(t_o)$, i.e., the current value of a (where t_o is the age of the universe), is scaled so that $a(t_o) = 1$, which gives the particular solution $a(t) = (t/t_o)^{2/3}$. Since there is a time when $a = 0$, there is a time with infinite density, i.e., a singularity. However, one may simply choose to have the universe start expanding from some $a(0) \neq 0$ to avoid the initial singularity.

But, wait a minute. Do the singularity theorems of Roger Penrose and Stephen Hawking not prove the initial singularity in Big Bang cosmology is unavoidable? Yes, given that a singularity is defined by past inextendible worldlines. That is, the spatial surface with $a(0) \neq 0$ would still be a singularity for these theorems even though the density of matter–energy is well defined there because we have worldlines beginning in space at that time. You can see why this definition of singularity is based on a dynamical bias, i.e., from a dynamical perspective there must be some *reason* for the initial condition of a worldline, and a worldline (with the space it is in) that just begins for no reason is dynamically unacceptable. But, the alternative is to extend worldlines (and space) into the past all the way to $a(0) = 0$ with its infinite density. So, dynamically speaking, the Big Bang is a serious problem, while for the all-at-once view a singularity defined by past inextendible worldlines without infinite density is a perfectly acceptable 'singularity.' That is, in the 4D adynamical view there is no concern with the origin of spacetime as a whole.

Second, this may be acceptable for avoiding the Big Bang singularity, but what about matter collapsing into a black hole? In that situation, the singularity is being approached from the past, so we cannot stop time-evolving a collapsing hypersurface in the context of the global spacetime. This is a context where discrete spacetime structure giving rise to the "small spatiotemporal relations" responsible for quanta of gravitational energy in the gravity double-slit experiment (mentioned above) might well obtain.

For example, there is already a discrete version of general relativity called *Regge calculus* [3, 12, 14, 30] where spacetime is modeled as a lattice of tiny spatiotemporally flat 4D cells called *simplices* (Figure 9B.1). Now suppose one models a collapsing star using a Schwarzschild vacuum surrounding a sphere of Einstein–de Sitter dust [12, pp. 851–853] and replaces the continuous dust model with its Regge calculus counterpart. The Regge equation for the dust-filled spatial hypersurfaces with continuous time is [24]

$$\frac{\pi - \cos^{-1}\left(\frac{v^2/c^2}{2(v^2/c^2+2)}\right) - 2\cos^{-1}\left(\frac{\sqrt{3v^2/c^2+4}}{2\sqrt{v^2/c^2+2}}\right)}{\sqrt{v^2/c^2+4}} = \frac{Gm}{2rc^2}, \qquad (9B.5)$$

where m is the mass on a graphical node, r is the spatial proper distance between nodes, and v is simply dr/dt. With $v^2/c^2 \ll 1$ an expansion of the left-hand side gives

$$\frac{v^2}{4c^2} + \mathcal{O}\left(\frac{v}{c}\right)^4 = \frac{Gm}{2rc^2}. \qquad (9B.6)$$

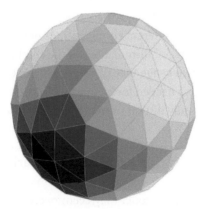

Fig. 9B.1 Simplices for a 2D sphere. Nodes are points on the triangles (simplices) and links are the sides of the triangles.

So, to leading order we have $v^2/2 = Gm/r$, which is just a Newtonian conservation-of-energy expression for a unit mass moving at escape velocity v at distance r from mass m. That is, in the case of Regge calculus we see that spacetime curvature enters as an angle between adjoining Newtonian simplices.

Interestingly, when one evolves this Regge calculus solution temporally such that the spatial hypersurface is shrinking, it 'freezes up' when adjacent nodes acquire a large relative velocity. This is called the *stop point problem* in Regge cosmology because the solution ceases to follow the continuous general relativity counterpart that does not stop shrinking towards the singularity. Obviously, if the continuous classical spacetime of general relativity becomes quantum in this context, then the stop point problem is not a problem, but precisely the solution expected of quantum gravity for avoiding singularities. This does not pose a threat to our quantum–classical contextuality any more than the analogous behavior of matter, e.g., this is totally analogous to the fact that atoms can be bodily objects in some contexts and quanta in others. The classical context needed for quantum–classical contextuality in this situation resides in the spacetime spatially distant from the high-density region.

We would also confront this situation in the early universe if the universe starts expanding from some spatial hypersurface with $a(0) \neq 0$ that is nonetheless very hot and dense. There is evidence of this in the Planck 2015 data for the cosmic microwave background angular power spectrum [1]. In such cases, we might expect classical bodily objects in continuous spacetime to be replaced by quantum fields in quantum spacetime. Accordingly, we might use Regge calculus to model a quantum spacetime region filled with quantum fields via the graphical version of quantum field theory called *lattice gauge theory*. In that case, the gauge fields of quantum field theory reside on the links of the simplices and the particle fields are scalars or vectors at each node. The transition amplitudes between different configurations of geometry and fields are then given by the path integral formulation for the combination of actions per Regge calculus and lattice gauge theory, respectively [20]. For the early universe, the classical context needed for quantum–classical contextuality resides in the spacetime temporally distant from the high-density region [11].

Contextuality, Dark Energy, and Dark Matter

While the Einstein–de Sitter model may seem overly simplistic, it only differs from the concordance model of cosmology in that it lacks a cosmological constant Λ, which was introduced to account for the apparent accelerating expansion of the universe inferred from the Supernova Cosmology Project

Union2.1 supernova type Ia data. However, one can fit this supernova data with the Einstein–de Sitter model as shown by Deur (above), or by using a simple perturbative correction to proper distance, no Λ and no accelerating expansion needed [23, 24]. The perturbative correction is motivated by the contextuality of mass in general relativity.

For example, suppose again that a Schwarzschild vacuum surrounds a sphere of FLRW dust in general, rather than just the flat, dust-filled model used above. The mass M of the surrounding Schwarzschild metric can be less than, greater than, or equal to the proper mass M_p measured locally inside the sphere of FLRW dust, depending on whether the FLRW model is spherical, hyperbolic, or flat, respectively [22]. We should quickly point out that this may prima facie seem to constitute a violation of the Principle of Equivalence, as understood to mean inertial mass equals gravitational mass, since inertial mass cannot be equal to two different values of gravitational mass. But, the Principle of Equivalence simply says that spacetime is locally flat [29, pp. 68–69] and that is certainly not being violated here, nor with any solution to Einstein's equations.

Here, one can identify a 'reference frame' with each of the different spatiotemporal geometries associated with one and the same matter source. That mass can have two different values is then attributed to the fact that G must have the same value in both of these reference frames (although G isn't necessarily a fundamental constant [9]), i.e., the contextuality of mass in general relativity can be said to follow from NPRF + G [25]. The contextuality of mass can then be used to solve the missing mass problem (the mystery of dark matter [26, 27]) by fitting THINGS data for galactic rotation curves [28], ROSAT/ASCA data for the mass profiles of X-ray clusters [4, 15, 16], and the Planck 2015 data for the cosmic microwave background angular power spectrum [1]. Conventional wisdom holds that all of these data point to the existence of dark matter. However, we fit all of these data as well as any competitor with a simple contextual correction to the mass of the ordinary (baryonic) matter present, no dark matter needed, just as Deur did (above).

In conclusion, the 4D contextuality proposed here greatly simplifies the problems of quantum gravity, singularities in general relativity, dark energy, and dark matter.

References

[1] P. A. R. Ade et al., *Planck 2013 results. XVI. Cosmological parameters*, Astronomy & Astrophysics, 571 (2014), p. A16.

[2] N. Ashby, *Relativity and the global positioning system*, Physics Today, 55 (2002), pp. 41–47.

[3] J. W. Barrett, *The geometry of classical Regge calculus*, Classical and Quantum Gravity, 4 (1987), pp. 1565–1576.

[4] J. Brownstein and J. Moffat, *Galaxy rotation curves without nonbaryonic dark matter*, The Astrophysical Journal, 636 (2006), pp. 721–741.

[5] S. Chen, *What Next for Gravitational Wave Detection?*, APS News, 28 (2019). https://www.aps.org/publications/apsnews/201906/wave.cfm.

[6] A. Deur, *Implications of graviton–graviton interaction to dark matter*, Physics Letters B, 676 (2009), pp. 21–24.

[7] A. Deur, *What can QCD teach us about Dark Matter?*, 2014. http://www.phys.virginia.edu/Files/fetch.asp?EXT=Seminars:2693:SlideShow.

[8] A. Deur, *An explanation for dark matter and dark energy consistent with the standard model of particle physics and general relativity*, The European Physical Journal C, 79 (2019), p. 883.

[9] T. Jacobson, *Thermodynamics of Spacetime: The Einstein Equation of State*, Physical Review Letters, 75 (1995), pp. 1260–1263.

[10] S. Lundeberg, *Astrophysics collaboration led by Oregon State finds "chorus" of gravitational waves*, Oregon State University Newsroom, 28 June (2023). https://today.oregonstate.edu/news/astrophysics-collaboration-led-oregon-state-finds-chorus-gravitational-waves.

[11] N. D. Mermin, *Making better sense of quantum mechanics*, Reports on Progress in Physics, 82 (2019), p. 012002.

[12] C. W. Misner, K. S. Thorne, and J. A. Wheeler, *Gravitation*, W. H. Freeman, San Francisco, 1973.

[13] PBS Space Time, *Quantum Gravity and the Hardest Problem in Physics*, 2018. https://www.youtube.com/watch?v=YNEBhwimJWs.

[14] T. Regge, *General relativity without coordinates*, Nuovo Cimento, 19 (1961), pp. 558–571.

[15] T. H. Reiprich, *Cosmological Implications and Physical Properties of an X-Ray Flux-Limited Sample of Galaxy Clusters*, PhD thesis, Ludwig-Maximilians-Universitaet, 2001.

[16] T. H. Reiprich and H. Böhringer, *The mass function of an X-ray flux-limited sample of galaxy clusters*, The Astrophysical Journal, 567 (2002), pp. 716–740.

[17] S. Rowan and J. Hough, *The detection of gravitational waves*, in 1998 European School of High-Energy Physics, 1998, pp. 301–311.

[18] Royal Swedish Academy of Sciences, *The Nobel Prize in Physics 2017*, 2017. https://www.nobelprize.org/prizes/physics/2017/press-release/.

[19] C. Sargent, A. Deur, and B. Terzic, *Hubble tension and gravitational self-interaction*, 2023. Preprint. https://arxiv.org/abs/2301.10861.

[20] M. Silberstein, W. M. Stuckey, and T. McDevitt, *Beyond the Dynamical Universe: Unifying Block Universe Physics and Time as Experienced*, Oxford University Press, Oxford, 2018.

[21] L. Sokolowski and A. Staruszkiewicz, *On the issue of gravitons*, Classical and Quantum Gravity, 23 (2006), pp. 5907–5917.

[22] W. M. Stuckey, *The observable universe inside a black hole*, American Journal of Physics, 62 (1994), pp. 788–795.

[23] W. M. Stuckey, T. McDevitt, and M. Silberstein, *Explaining the Supernova Data without Accelerating Expansion*, International Journal of Modern Physics D, 21 (2012), p. 1242021.

[24] W. M. Stuckey, T. McDevitt, and M. Silberstein, *Modified Regge calculus as an explanation of dark energy*, Classical and Quantum Gravity, 29 (2012), p. 055015.

[25] W. M. Stuckey, T. McDevitt, and M. Silberstein, *"Mysteries" of Modern Physics and the Fundamental Constants c, h, and G*, Quanta, 11 (2022), pp. 5–14.

[26] W. M. Stuckey, T. McDevitt, A. K. Sten, and M. Silberstein, *End of a Dark Age?*, International Journal of Modern Physics D, 25 (2016), p. 1644004.

[27] W. M. Stuckey, T. McDevitt, A. K. Sten, and M. Silberstein, *The Missing Mass Problem as a Manifestation of GR Contextuality*, International Journal of Modern Physics D, 27 (2018), p. 1847018.

[28] F. Walter, E. Brinks, W. J. G. de Blok, F. Bigiel, R. C. K. Jr, M. D. Thornley, and A. Leroy, *THINGS: The H I Nearby Galaxy Survey*, The Astronomical Journal, 136 (2008), pp. 2563–2647.

[29] S. Weinberg, *The cosmological constant problems (talk given at Dark Matter 2000, February, 2000)*, 2000. Preprint. https://arxiv.org/abs/astro-ph/0005265.

[30] R. M. Williams and P. A. Tuckey, *Regge calculus: A brief review and bibliography*, Classical and Quantum Gravity, 9 (1992), pp. 1409–1422.

9C

The Delayed-Choice Experiment

Xiao-song Ma, Johannes Kofler, and Anton Zeilinger note that the idea of a delayed-choice experiment can be traced back to Heisenberg in 1927 [5]. Wheeler introduced a delayed-choice experiment in 1978 and Zeilinger carried out a version of it in 2004 (Figure 9C.1). The mystery of this experimental outcome (Figure 9C.2) does not require any familiarity with the formalism of quantum mechanics or knowledge of what Zeilinger's nomenclature for his outcomes even means. All you need to recognize is that the two different experimental configurations in Figure 9C.1 produce the two different experimental outcomes in Figure 9C.2. And, that happens because a laser beamed into a crystal produces a pair of entangled photons (Figure 9C.1). In this case, measurement configurations for the photons at detector D1 are correlated

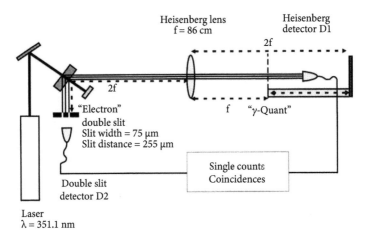

Fig. 9C.1 Zeilinger's delayed-choice experiment. Anton Zeilinger, "Why the quantum? 'It' from 'bit'? A participatory universe? Three far-reaching challenges from John Archibald Wheeler and their relation to experiment," in *Science and Ultimate Reality: Quantum Theory, Cosmology and Complexity*, John D. Barrow, Paul C. W. Davies, and Charles L. Harper, Jr. (eds.), Cambridge University Press, Cambridge, 2004, pp. 201–220. Reproduced by permission of Cambridge University Press.

Einstein's Entanglement. W. M. Stuckey, Michael Silberstein, and Timothy McDevitt, Oxford University Press.
© Oxford University Press (2024). DOI: 10.1093/9780198919698.003.0013

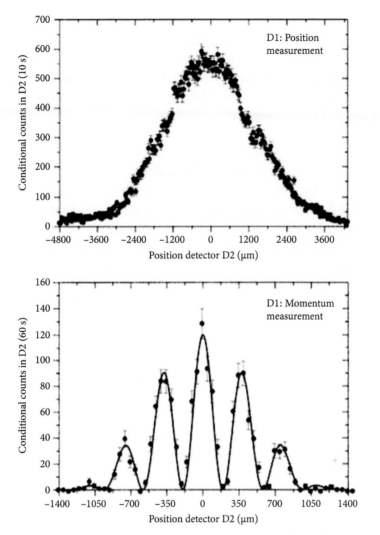

Fig. 9C.2 Outcomes in Zeilinger's delayed-choice experiment. Zeilinger, "Why the quantum?" Reproduced by permission of Cambridge University Press.

with the measurement outcomes for their entangled partners at detector D2, as follows.

The experimentalist can decide whether to locate D1 (shown under the words "Heisenberg detector D1" in Figure 9C.1) at a distance of one focal length (f, the "momentum measurement" labeled "γ-Quant" in Figure 9C.1) or two focal lengths (2f, the "position measurement," the position farthest to the right from the "Heisenberg lens" in Figure 9C.1) behind the "Heisenberg lens" in Figure 9C.1. The photons detected at D2 (shown to the left of the words "double slit detector" and immediately above the words "Double slit detector D2") produce two different patterns corresponding to the two locations for D1, i.e., the upper (particle) pattern (labeled "D1: Position measurement") and the lower (interference/wave) pattern (labeled "D1: Momentum measurement") of Figure 9C.2.

Delayed Choice?

Notice that the distance from the photon source to D2 is much shorter than the distance from the photon source to either location for D1. The two entangled photons are emitted at the same time from the same place traveling at the same speed, so the first photon to reach a detector is at D2. But, the photon's behavior at D2 is supposedly determined by the experimentalist's choice of where to locate D1. How does the photon at D2 know where D1 will be placed when that placement can occur in the future of the photon detection at D2? Does the experimentalist not have a real (delayed) choice about where to place D1 that can be made after the photon has been detected at D2 and before its entangled partner has been detected at D1?

Before we continue, we should point out that *any particular outcome* is consistent with either pattern in Figure 9C.2, so the mysterious 'delayed choice' nature of the result is only evident in the aggregate. Of course, while we cannot attribute the mystery to each detection event per se, they do add up to that mysterious aggregate, so the mystery still obtains. In short, it is a mystery associated with the aggregate rather than any particular detection outcome.

If causal influences must proceed from past to future and the correlated outcomes at D1 and D2 are causally related (as required by Reichenbach's Principle), then we have two choices. Either the photon at D2 causally determines the experimentalist's choice of where to put D1 (however that choice is made), or there is some superdeterministic common cause for both outcomes. We discussed the difficulty faced by the second option, and that difficulty is also true for the first option.

That is, as we stated in Chapter 3, the dynamical law or mechanistic causal process responsible for the placement of D1 (stemming either from the detection event at D2 or superdeterministically from some common event in the past of both outcomes) would have to account for any means of that placement. For example, the placement of the detector D1 might be decided by a random number generator in a computer chip, monkeys throwing darts, human choices, the polarization of light emitted from a quasar billions of years ago, etc. That does not mean there can be no such constructive account (e.g., see Palmer's approach [6, 7]), but it is going to be complicated and Bell's theorem tells us it will violate statistical independence in 'conspiratorial fashion' [8].

Alternatively, we could abandon the requirement that causes must always precede their effects in time, i.e., we could adopt retrocausality. As we explained in Chapters 3 and 5, one might relax the strict requirement that all causal influences proceed from past to future as in classical physics and accept time-symmetric explanation. Again, comparatively speaking, this type of explanation is best thought of as atemporal or all-at-once explanation. Price, Liu, and Wharton explicate this idea in terms of retrocausality [9, 11]; Hance, Hossenfelder, and Palmer go for the idea of superdeterminism [3]; and Adlam and Rovelli invoke relational quantum mechanics [1].

The goal for a constructive approach to retrocausality would be to find some spatiotemporally constructive mechanism responsible for coordinating the outcomes at D1 and D2, for example. Again, such a constructive explanation is perhaps not impossible [10], but it is going to be complicated and Bell's theorem tells us it will violate statistical independence.

That means, if we do not want to violate statistical independence, intersubjective agreement, or unique experimental outcomes, we will have to abandon constructive attempts to explain the delayed-choice experiment. That is perhaps not surprising, since the experiment is a particular exploitation of entanglement. So, let's see how this experiment is explained in principle fashion by Information Invariance & Continuity as justified by the relativity principle, i.e., NPRF + h.

Outcomes Explained

While Zeilinger only conducted the two complementary measurements of momentum p and position x, he has a continuum of choices (superpositions of x and p measurements) for the location of D1 between those two extremes, just as we had with orientations of the SG magnets between the z axis and the x axis for complementary spin measurements and with alignments of the polarizer between the vertical and diagonal axes for complementary polarization measurements. In the spin and polarization cases, we had two (binary) outcomes for each spin or polarization (qubit) measurement, but at first glance it would appear that we are missing the binary outcomes for the Zeilinger experiment.

However, Zeilinger's experiment can also be understood via binary outcomes corresponding to the qubit for the double-slit experiment.

Qubits in the Double-Slit Experiment

In general, we can model the double-slit experiment via qubits as shown in Figure 9C.3. If we do an x measurement on the state $|\psi\rangle$ for the double-slit experiment, we have two possible outcomes, i.e., the quantum 'went through' the left slit or the quantum 'went through' the right slit. If the first outcome is realized, then $|\psi\rangle \rightarrow |\text{LS Yes}\rangle$. If the second outcome is realized, then

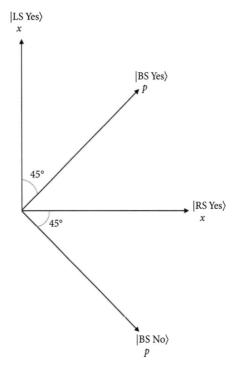

Fig. 9C.3 Hilbert space qubits for the double-slit experiment. The basis for a position measurement x is left slit "yes," $|\text{LS Yes}\rangle$, and right slit "yes," $|\text{RS Yes}\rangle$. Obviously, left slit "yes" is right slit "no" and vice versa. The complementary momentum measurement p is both slits "yes," $|\text{BS Yes}\rangle$ (constructive interference), and both slits "no," $|\text{BS No}\rangle$ (destructive interference). The angle between the two bases is 45°, just like the bases for the complementary spins S_z and S_x (Figure 6.4) and complementary photon polarizations V and D (Figure 9.1).

$|\psi\rangle \to |RS\ Yes\rangle$. The double-slit pattern produced by these states (upper pattern in Figure 9C.2) does not have a constructive interference maximum or a destructive interference minimum (lower pattern in Figure 9C.2) that can be used to determine the wavelength λ (Chapter 4) for the quantum, and no wavelength means the momentum $p = h/\lambda$ cannot be determined at all.[1]

Conversely, if we do a p measurement on the state $|\psi\rangle$ for the double-slit experiment, we have two possible outcomes, i.e., a bright strip for constructive interference (a maximum) or a dark strip for destructive interference (a minimum). If the first outcome is realized, then $|\psi\rangle \to |BS\ Yes\rangle$. If the second outcome is realized, then $|\psi\rangle \to |BS\ No\rangle$. Again, the maxima (bright strips) and minima (dark strips) of an interference pattern allow a definite wavelength to be calculated, so the quantum's momentum is known exactly while the position is as ambiguous as can be.

That is, the probability that the quantum 'went through' the left slit for $|BS\ Yes\rangle$ is $|\langle LS\ Yes|BS\ Yes\rangle|^2 = 0.5\ (50\%)$ and the probability that the quantum 'went through' the right slit for $|BS\ Yes\rangle$ is $|\langle RS\ Yes|BS\ Yes\rangle|^2 = 0.5\ (50\%)$. The same holds for $|BS\ No\rangle$. That means the bright and dark strips (momentum measurement outcomes) provide exact information for the wavelength of the quantum (and therefore momentum p), but are totally ambiguous with respect to which slit the quantum 'passed through' (position x). Of course, these calculations also show us that the converse is true, i.e., $|\langle RS\ Yes|BS\ Yes\rangle|^2 = 0.5$ (50%) and $|\langle RS\ Yes|BS\ No\rangle|^2 = 0.5\ (50\%)$ (same for $|LS\ Yes\rangle$) means knowing x makes the p measurement outcome as uncertain as possible.

Solving "The Only Mystery" of Quantum Mechanics

Feynman famously said that the mystery of wave–particle duality in the double-slit experiment is [2, Chapter 37]:

a phenomenon which is impossible, *absolutely* impossible to explain in any classical way, and which has in it the heart of quantum mechanics. In reality, it contains the *only* mystery. We cannot make the mystery go away by 'explaining' how it works. We will just *tell* you how it works. In telling you how it works we will have told you about the basic peculiarities of all quantum mechanics.

---

[1] This can be related to the uncertainty relation for position and momentum in interference experiments [2, Chapter 38], but we are not interested in that here.

Also, the manner by which one would obtain the two 'which slit' outcomes from the "position measurement" pattern in Figure 9C.2 is not obvious in Zeilinger's version of this experiment, but that does not concern us here. If you are interested, see Kim et al.'s version of the experiment [4].

We can solve "the *only* mystery" of quantum mechanics per NPRF + h, as we did in the Stern–Gerlach and polarization experiments.

Recall for the spin-$\frac{1}{2}$ particle, Information Invariance & Continuity entails that everyone measures the same value for h regardless of the orientation of their SG magnets, i.e., regardless of the spatial orientation of their inertial reference frames. Since we are talking about a constant of Nature having the same value in all inertial reference frames (related in this case by spatial rotations), this empirically discovered fact can be justified by NPRF. To see how the classical formalism follows on average from the quantum formalism, we showed how the classical constructive account predicts that atomic dipoles along $\hat{z}$ should be deflected by $\cos(\theta)$ when they pass through SG magnets oriented at $\hat{b}$ making an angle θ with $\hat{z}$. Instead, we find only full deflections of ± 1 ($\pm \hbar/2$) along $\hat{b}$ that *average* to $\cos(\theta)$. This gives you 'average-only' projection of spin angular momentum between these reference frames as a consequence of NPRF + h.

For polarization, Information Invariance & Continuity entails that everyone measures the same value for h regardless of the orientation of their polarizing filter, i.e., regardless of the spatial orientation of their inertial reference frames, which is easily justified by NPRF just like the spin-$\frac{1}{2}$ particle. The empirically discovered fact in this case is that a photon either passes or does not pass through a polarizing filter, i.e., there are no 'partial photons' passed by a polarizer. We can say that our empirically discovered fact in this case results because $E = hf$ and if a fractional photon passed through a polarizer, the value of h would be reduced by that fraction, so you would have two inertial reference frames related by a spatial rotation whereby h had two different values in violation of NPRF. This gives you 'average-only' transmission of polarized electromagnetic energy between these reference frames as a consequence of NPRF + h. Now let's provide a similar NPRF + h solution to the mystery of wave–particle duality in the double-slit experiment.

Our measurement context will be the interference pattern of a momentum measurement as in the Zeilinger setup, so our qubit is that of Figure 9C.4 with outcomes $|\text{BS Yes}\rangle$ of total constructive interference at a maximum and $|\text{BS No}\rangle$ of total destructive interference at a minimum. Our state is $|\psi\rangle = |\text{BS Yes}\rangle$ and the measurement in this context will be a spatially localized detector moved a distance x along the interference pattern from the central maximum toward a minimum on either side (Figure 9C.5 or 9C.6). The amount of momentum it records at each location along the interference pattern depends on how much constructive interference exists at that location, so

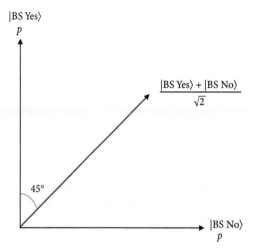

Fig. 9C.4 Hilbert space qubit for momentum measurement in an interference pattern for the double-slit experiment. The vector $(|\text{BS Yes}\rangle + |\text{BS No}\rangle)/\sqrt{2}$ corresponds to that location between the central maximum ($|\text{BS Yes}\rangle$) and first adjacent minimum ($|\text{BS No}\rangle$) where the electric field is $1/\sqrt{2}$ of that at the central maximum per the classical wave analysis.

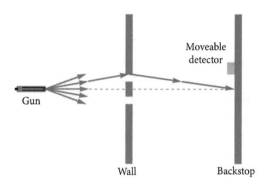

Fig. 9C.5 Bullets used in the double-slit experiment per Feynman. The measured pattern of bullets would be that of the position measurement / upper pattern in Zeilinger's Figure 9C.2.

it will start from a maximum and gradually diminish to zero at the adjacent minimum where there is no constructive interference for the state $|\psi\rangle$.

For example, with $|\psi\rangle = |\text{BS Yes}\rangle$ the electric fields $E_1 = E_2$ of the electromagnetic waves from the two slits add in phase at the central maximum to E_o (say) and we have total constructive interference, while they are totally out of phase and add to zero at the adjacent minimum. At any point between the

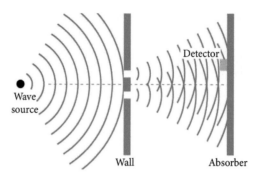

Fig. 9C.6 Water waves used in the double-slit experiment per Feynman. The measured pattern of wave intensity would be that of the momentum measurement / lower pattern in Zeilinger's Figure 9C.2.

central maximum and adjacent minimum the electric field is given by $E_o \cos \phi$, where $\phi = (\pi d \sin \theta)/\lambda$ with the angle θ and slit separation d as shown for the two slits per rays 1 and 2 in Figure 9C.7. So, when $\phi = \pi/4$ the electric field is $E_o/\sqrt{2}$ and the electromagnetic wave intensity (proportional to the wave momentum per the Poynting vector) is half that at the maximum. This will correspond to the Hilbert space vector $(|\text{BS Yes}\rangle + |\text{BS No}\rangle)/\sqrt{2}$ in the quantum case (Figure 9C.4).

Mechanically speaking, this is exactly how Zeilinger's delayed-choice experiment worked; his detector recorded counts over a 60-second time interval at each location along the interference pattern as shown in the lower pattern of Figure 9C.2 (see also Feynman's versions of the double-slit experiment with bullets, water waves, and electrons [2, Chapter 37]). A classical wave measurement of this sort would obtain momentum in a continuum sense (like the water waves in Figure 9C.6), rather than discrete detector clicks (like the bullets in Figure 9C.5), at each spatial location. Of course, each detector click represents a quantum of momentum/energy, so counting detector clicks is still a measurement of momentum (over the area of the detector where each click delineates a unit of time).

As with the polarization experiment, we then reduce the rate at which momentum is emitted by the source so that only one quantum of momentum $p = h/\lambda$ is 'passing through' the slits at a time. The classically continuous wave description says p will be distributed spatially along the detector, so our spatially localized momentum measurement described above will measure a fraction of the momentum $p = h/\lambda$ at any given location x, which means h is reduced by precisely that same fraction in the inertial reference frame at x. Since fractional measurements of h are forbidden by NPRF (our inertial reference frames here are related by spatial translations in x), our momentum measurements at each point x instead obtain a number of quanta (each with the same value for h in $p = h/\lambda$) per fixed time (60 seconds in Zeilinger's experiment) relative to the central maximum and that collection of quantum measurement results maps to the continuum distribution of momentum along the interference pattern predicted by the classical wave interference analysis on average.

Our explanation of the double-slit experiment per NPRF + h and the qubit shows that Feynman was essentially correct. The mystery of the double-slit experiment resides in the fact that NPRF + h demands we obtain quanta of momentum $p = h/\lambda$ in the inertial reference frames associated with the locations x (as with the bullets in Figure 9C.5), rather than a continuum of wave intensity (as with the water waves in Figure 9C.6). However, the quantum measurement context is that for momentum p, so an interference pattern allowing us to compute λ for use in $p = h/\lambda$ is required. Therefore, the interference pattern per classical wave mechanics is what must obtain on average in this quantum-mechanical context as in the momentum measurement / lower pattern in Zeilinger's Figure 9C.2. In all three examples, NPRF + h demands that a classically continuous quantity (angular momentum, energy, momentum) be quantized, so that you have to take an average to recover the classical continuity.

That summarizes how the qubits of Figure 9C.3 map to the outcomes of the double-slit experiment in general (Figure 9C.2). As we promised in Chapter 6, we will now show you how these qubits are used to generate higher-dimensional Hilbert spaces for the triple-slit, quadruple-slit, ... experiments per the information-theoretic reconstructions. We will do this for the triple-slit experiment explicitly and (hopefully) this will give you a sense of how one would build the Hilbert spaces for the quadruple- and higher-slit experiments from there.

Qubits in the Triple-Slit Experiment

In the triple-slit experiment, the x measurement has three possible outcomes, call them 1 (top slit, ray 1), 2 (middle slit, ray 2) and 3 (bottom slit, ray 3) in Figure 9C.7. You can view these as a collection of double-slit qubits, i.e., slits 1 and 2, slits 2 and 3, and slits 1 and 3. These appear in the Hilbert space for the triple-slit experiment (Figure 9C.8) as the pairs ($|1\ \text{Yes}\rangle, |2\ \text{Yes}\rangle$), ($|2\ \text{Yes}\rangle, |3\ \text{Yes}\rangle$), and ($|1\ \text{Yes}\rangle, |3\ \text{Yes}\rangle$).

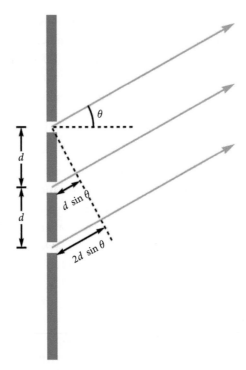

Fig. 9C.7 The triple-slit experiment.

The p measurement likewise has three possible outcomes associated with the bright strip of the principle constructive interference maxima and the two dark strips of destructive interference minima between any pair of principle maxima (the tall peaks at 0 and $\pm\lambda/d$ in Figure 9C.9). Notice that in Figure 9C.7 all slits will contribute to the principle maxima, so the associated momentum state corresponds to $|BS\ Yes\rangle$ in the qubit momentum Hilbert space.

To see that for the principle maxima look again at Figure 9C.7. In the direction where $d\sin\theta = \lambda$ we see that rays 1 and 2 are in phase. In this case, $2d\sin\theta = 2\lambda$ so rays 2 and 3 are also in phase, which means all three rays are in phase in this direction. This p measurement outcome corresponds to the qubit state $|BS\ Yes\rangle$, so we labeled it as slits 1, 2, 3 "yes," $|123\ Yes\rangle$ (Figure 9C.8). Notice again that this state is maximally vague as to which slit (x measurement outcome) each quantum 'passed through' to contribute to the principle maximum in that direction, i.e., $|\langle X\ Yes|123\ Yes\rangle|^2 = \frac{1}{3}$ for $X = 1, 2,$ or 3.

As for the dark strips of destructive minima, we need rays 1, 2, and 3 to all cancel in some direction θ for d and λ. Let us find the two minima between the central principle maximum and the next principle maximum to the right

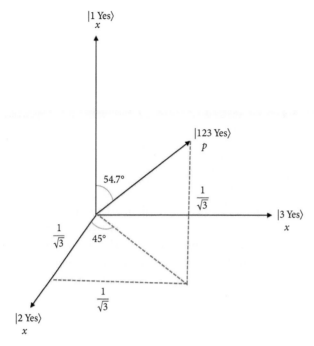

Fig. 9C.8 Hilbert space for the triple-slit experiment. There are three x measurement outcome states, $|1\text{ Yes}\rangle$, $|2\text{ Yes}\rangle$, and $|3\text{ Yes}\rangle$, corresponding to the three slits in Figure 9C.7. We have included the p measurement outcome state corresponding to the principle maxima
$|123\text{ Yes}\rangle = (|1\text{ Yes}\rangle + |2\text{ Yes}\rangle + |3\text{ Yes}\rangle)/\sqrt{3}$.

in Figure 9C.9. Those are located in the direction where $d\sin\theta = \lambda/3$ (the minimum next to the central principle maximum) and in the direction where $d\sin\theta = 2\lambda/3$ (the minimum next to the second principle maximum). In these cases, parts of rays 2 and 3 cancel ray 1 and the other parts of rays 2 and 3 cancel each other. The two Hilbert space vectors for these two states are a bit complicated, but we show them so you can see how it all works.

Let us label these states $|123\text{a No}\rangle$ and $|123\text{b No}\rangle$ in analogy with $|BS\text{ No}\rangle$ for the double-slit minimum. We have:

$$|123\text{a No}\rangle = \frac{1}{\sqrt{3}}|1\text{ Yes}\rangle + \left(\frac{-1}{2\sqrt{3}} + \frac{i}{2}\right)|2\text{ Yes}\rangle + \left(\frac{-1}{2\sqrt{3}} - \frac{i}{2}\right)|3\text{ Yes}\rangle,$$

$$|123\text{b No}\rangle = \frac{-1}{\sqrt{3}}|1\text{ Yes}\rangle + \left(\frac{1}{2\sqrt{3}} + \frac{i}{2}\right)|2\text{ Yes}\rangle + \left(\frac{1}{2\sqrt{3}} - \frac{i}{2}\right)|3\text{ Yes}\rangle.$$

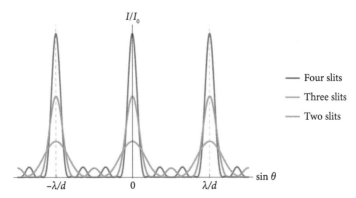

Fig. 9C.9 Double-, triple-, and quadruple-slit interference patterns. Note the locations of the destructive interference minima relative to the principle and secondary maxima. We have one minima for two slits, two minima for three slits, and three minima for four slits. These correspond to the momentum measurement states spanning the momentum Hilbert space orthogonal to the state corresponding to the principle maximum, i.e., $|\text{BS Yes}\rangle$, $|123\ \text{Yes}\rangle$, and $|1234\ \text{Yes}\rangle$, respectively.

You can check that the set of momentum states $|123\ \text{Yes}\rangle$, $|123\text{a No}\rangle$, and $|123\text{b No}\rangle$ forms an orthonormal set like $\hat{x}$, $\hat{y}$, and $\hat{z}$ in real space (you just have to take the complex conjugate when flipping a "ket" $|X\rangle$ to a "bra" $\langle X|$ for doing a vector dot product). Notice how for both $|123\text{a No}\rangle$ and $|123\text{b No}\rangle$ the real parts of $|2\ \text{Yes}\rangle$ and $|3\ \text{Yes}\rangle$ add to -1 times the $|1\ \text{Yes}\rangle$ part, and the imaginary parts of $|2\ \text{Yes}\rangle$ and $|3\ \text{Yes}\rangle$ are the negatives of each other. This corresponds to the fact that parts of rays 2 and 3 cancel ray 1 and the other parts of rays 2 and 3 cancel each other, as we said above. This is the momentum qubit structure at work in the triple-slit experiment. Finally, we point out that $|\langle X\ \text{Yes}|123\text{a No}\rangle|^2 = \frac{1}{3}$ and $|\langle X\ \text{Yes}|123\text{b No}\rangle|^2 = \frac{1}{3}$ for $X = 1, 2$, or 3. So, again, these states are maximally vague as to which slit (x measurement outcome) contributed to these minima. By the way, you can see why complex numbers are essential in making all of this happen.

Hopefully, it is apparent how to build upon this structure to produce the Hilbert space for four slits, i.e., you have your principle maxima represented by $|1234\ \text{Yes}\rangle$ and the three-dimensional minima subspace (associated with the three minima between any pair of principle maxima, see Figure 9C.9) given by a Hilbert space like Figure 9C.10 per the second axiom for Dakić and Brukner's reconstruction (Chapter 6):

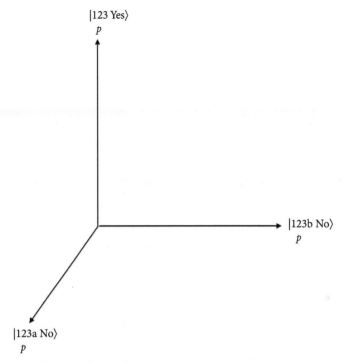

Fig. 9C.10 Momentum Hilbert space for the triple-slit experiment. There are three p measurement outcome states, i.e., $|123\,\text{Yes}\rangle$ for the principle constructive maxima and two orthogonal states in the plane perpendicular to $|123\,\text{Yes}\rangle$ for the destructive minima (Figure 9C.9), i.e., $|123a\,\text{No}\rangle$ and $|123b\,\text{No}\rangle$.

(2) (Locality) The state of a composite system is completely determined by local measurements on its subsystems and their correlations.

Keep in mind that this is just one particular physical example of how the Hilbert space formalism of quantum mechanics is built from the qubit. The information-theoretic reconstructions are general, so they hold for any such physical implementation.

Contextuality Not Causality

Now that we see how the double-slit experiment is modeled with qubits, we can solve the mystery of Zeilinger's delayed-choice experiment via NPRF + h. Since the location of D1 is being used to establish the corresponding entangled qubit measurement at D2 (x or p or some superposition thereof), we are not recording binary outcomes for the D1 measurement, so our states look like

$$|\psi\rangle = \frac{|\text{RS Yes}\rangle \otimes |\text{D1 at 2f}\rangle + |\text{LS Yes}\rangle \otimes |\text{D1 at 2f}\rangle}{\sqrt{2}}$$

or

$$|\psi\rangle = \frac{|\text{BS Yes}\rangle \otimes |\text{D1 at f}\rangle + |\text{BS No}\rangle \otimes |\text{D1 at f}\rangle}{\sqrt{2}}$$

or some superposition thereof. Notice that $|\psi\rangle$ does not represent some 'thing' with intrinsic properties, rather it provides a distribution of measurement outcomes for a specific experimental configuration, i.e., it is contextual by its very construct.

As we pointed out in Chapter 3, the commutator for these complementary measurements is given by Planck's constant h, so we are getting binary outcomes at D2 precisely because h is fixed at a nonzero value. Again, in information-theoretic terms that means there is a limit on the simultaneous information one can have for complementary variables. In spin angular momentum or polarization measurements the invariance of h obtains under spatial rotation (SG magnets rotate from 0° to 90° or polarizer rotates from 0° to 45°) while in the position–momentum measurements for the double-slit experiment the invariance of h obtains under spatial translation (D1 translates from 2f to f). In all three cases, we are talking about transformations between inertial reference frames.

In this sense, Information Invariance & Continuity as justified by the relativity principle (NPRF + h) solves the mystery of Zeilinger's delayed-choice experiment. In general, we are saying that the way to understand the outcomes at D2 in relation to the position of D1 has nothing to do with causality. NPRF + h is a 4D (all-at-once) AGC that constrains the relationship between the outcomes at D2 and the position of D1. So, it is neither the case that the experimentalist's choice of where to locate D1 'caused' the outcomes at D2 nor the case that the outcomes at D2 'caused' the experimentalist to choose a particular location for D1. The mystery of the delayed-choice experiment is just another example of misplaced constructive bias.

References

[1] E. Adlam and C. Rovelli, *Information is Physical: Cross-Perspective Links in Relational Quantum Mechanics*, 2022. Preprint. https://arxiv.org/abs/2203.13342.

[2] R. Feynman, R. Leighton, and M. Sands, *The Feynman Lectures on Physics Volume 1*, 1963. https://www.feynmanlectures.caltech.edu/I_toc.html.

[3] J. Hance, S. Hossenfelder, and T. Palmer, *Supermeasured: Violating Bell-Statistical Independence without Violating Physical Statistical Independence*, Foundations of Physics, 52 (2022), p. 81.

[4] Y. Kim, R. Yu, S. P. Kulik, Y. H. Shih, and M. O. Scully, *Delayed Choice Quantum Eraser*, Physical Review Letters, 84 (2000), pp. 1–5.

[5] X.-S. Ma, J. Kofler, and A. Zeilinger, *Delayed-choice gedanken experiments and their realizations*, Reviews of Modern Physics, 88 (2016), p. 015005.

[6] T. Palmer, *Discretised Hilbert space and superdeterminism*, 2022. Preprint. https://arxiv.org/abs/2204.05763.

[7] T. Palmer, *Superdeterminism Without Conspiracy*, 2023. https://arxiv.org/abs/2308.11262.

[8] T. Palmer and C. Timpson, *Superdeterminism and No-Conspiracy Revisited: A Debate*, 2023. https://www.youtube.com/watch? v=y_GtgyGjzPU.

[9] K. Wharton, *The universe is not a computer*, in Questioning the Foundations of Physics, A. Aguirre, B. Foster, and Z. Merali, eds., Springer, Heidelberg, 2015, pp. 177–190.

[10] K. Wharton and N. Argaman, *Bell's theorem and locally mediated reformulations of quantum mechanics*, Reviews of Modern Physics, 92 (2020), p. 021002.

[11] K. Wharton and R. Liu, *Entanglement and the Path Integral*, 2022. Preprint. https://arxiv.org/abs/2206.02945.

9D

Schrodinger's Cat and Wigner's Friend

Early on, the idea of superposition struck some physicists as absurd and in 1935 Schrödinger wrote [10], "One can even set up quite ridiculous cases." To make his point, he assumed a cat was closed out of sight in a box with radioactive material that would decay with 50% probability within an hour. If a radioactive decay occurred, a deadly gas would be released in the box, killing the cat. Since the decay was represented by a quantum wavefunction in a superposition of 50% "yes" and 50% "no" regarding the decay after one hour, the cat was also represented by a quantum wavefunction in a superposition of 50% "alive" and 50% "dead" (Figure 9D.1). Schrödinger wrote [10]:

> The wavefunction of the entire system would express this by having in it the living and dead cat (pardon the expression) mixed or smeared out in equal parts.

This has become known as Schrödinger's Cat.

Fig. 9D.1 Schrödinger's Cat.

Einstein's Entanglement. W. M. Stuckey, Michael Silberstein, and Timothy McDevitt, Oxford University Press.
© Oxford University Press (2024). DOI: 10.1093/9780198919698.003.0014

We discussed our all-at-once solution to the measurement problem, with which the Schrödinger's Cat thought experiment is most closely associated, in Chapter 3. In 1961, Eugene Wigner came up with a similar but potentially even weirder thought experiment [14]. He supposed his friend was hidden in a room and was going to make a measurement on a qubit with two equally possible outcomes, +1 and –1. For his friend in the room, there was a definite outcome of the measurement, but until Wigner opens the door to the room, i.e., makes his measurement, his friend and his friend's outcome are in a superposition state of 50% (Friend)(+1) plus 50% (Friend)(–1) (Figure 9D.2).

That means there are four possible outcomes for Wigner and his friend, half of which (outcomes 2 and 3) result in a violation of intersubjective agreement:

1. His friend's outcome was +1 and Wigner's outcome was (Friend)(+1).
2. His friend's outcome was +1 and Wigner's outcome was (Friend)(–1).
3. His friend's outcome was –1 and Wigner's outcome was (Friend)(+1).
4. His friend's outcome was –1 and Wigner's outcome was (Friend)(–1).

Wigner was actually arguing for the effect of consciousness on quantum measurement outcomes, but observers need not be conscious to make the point, and we will appropriate the Wigner's Friend experiment for our own purposes here (as presented in Subchapter 9A). There are more complicated versions involving two "Wigners" and two "friends" [3, 4, 5, 6], but we can keep things simple here and still make our point. So, what does our principle account of quantum mechanics say about Schrödinger's Cat and Wigner's Friend?

Most deeply, in both cases there is a confusion between bodily objects and quanta. As we explained above, atoms can be quanta in some contexts, e.g., silver atoms in the SG experiment, and they can be bodily objects in other contexts, e.g., rubidium atoms in an atomic trap. It is impossible to

Fig. 9D.2 Wigner's Friend.

establish simultaneous values for the noncommuting variables of quanta while bodily objects exchange enough quanta with other bodily objects to establish simultaneous values for their commuting variables.

In the case of Schrödinger's Cat, the cat is obviously a bodily object when placed in the box, so what changes when we close the lid? When we close the lid, we have stopped the cat from exchanging quanta of visible photons with our eyes, but does that mean the cat is no longer a bodily object? Of course not, the cat is still exchanging photonic quanta with the box and the box is exchanging photonic quanta with us, so the combination of observer, cat, box, and vial of gas still constitutes a system of interacting bodily objects, i.e., a subset of our objective spacetime model of reality. Is there a quantum component to this scenario at all? Of course there is: the radioactive source either decays or does not according to its wavefunction. The bodily object constituting the measurement device for that wavefunction is the mechanism that releases the poisonous gas (or not). As Jeffrey Bub points out [4], mere ignorance about classical information does not constitute a quantum system. Indeed, if Schrödinger's Cat is a qubit and not merely a classical bit, then there must be a measurement complementary to "open the box" where the outcome is the superposition of Live Cat + Dead Cat [13]. What would that be?

In the case of Wigner's Friend, we would need the friend and her entire lab to become a quantum for Wigner's measurement.[1] That means the friend and her lab cannot be exchanging quanta with Wigner and the rest of the universe, because Wigner and the rest of the universe constitute the collection of bodily objects exchanging enough quanta to establish the global spacetime reference frame containing all the personal reference frames for all observers and their data collection devices. This global spacetime model obtained via intersubjective agreement among all observers is required to make sense of the quantum-mechanical formalism, as we explained above. Every 'thing' in that global spacetime is either the worldline/worldtube for a bodily object or a spacetime region of action h for a quantum in the worldtube of a bodily object (subject to the caveat above about the complementary variables needed to construct an action).

If Wigner is merely opening the lab door (or not), he is conducting a classical measurement, as with the two buttons for our classical gloves experiment in Chapter 0 (Figure 0.1) or the box containing a ball in Chapter 6. By opening the door to the lab, Wigner would immediately acquire vast amounts of information via the exchange of many photonic quanta. This would allow him

---

[1] What follows here is a qualitative overview of [12]; see that reference for our detailed treatment of Wigner's Friend.

to ascertain the simultaneous values for many variables associated with the lab that would be noncommuting variables for a quantum, i.e., position and momentum. So, what would a quantum measurement look like for Wigner's Friend?

Well, it has to be possible for Wigner to measure superpositions of (Friend)(+1) and (Friend)(−1). For example, Wigner could make a measurement where 10% (Friend)(+1) plus 90% (Friend)(−1) is the final state for the outcome of "yes." How would Wigner make such a measurement? We have no idea, but supposing we could make sense of such a measurement, what would the outcome of such a measurement mean? Does the friend now exist with a memory of 10% (+1) plus 90% (−1)? That does not agree with quantum mechanics because the friend's single qubit measurement must produce either +1 or −1 according to quantum mechanics (assuming unique experimental outcomes).

In order for this to produce self-consistent outcomes for Wigner and his friend, as required for intersubjective agreement producing the objective spacetime model of reality, it must be the case that Wigner's measurement outcome of his friend and her outcome is consistent with what she says was her outcome. If Wigner's measurement corresponds to a direction in space (as with the orientation of SG magnets or the alignment of polarizers), it might be the case (however unlikely) that Wigner's measurement direction always corresponds to his friend's measurement direction as revealed once the friend and her lab start exchanging quanta with Wigner and the rest of the universe, which is necessary to establish common directions in space [2].

In other words, Wigner cannot know whether or not he is measuring a superposition of (Friend)(+1) and (Friend)(−1) unless he knows what her outcomes correspond to in terms of an experimental configuration in the global spacetime context of the objective spacetime model of reality, e.g., the orientation of SG magnets or the alignment of polarizers. But, he cannot know that until his friend and her lab start exchanging enough quanta with him and the rest of the universe to establish herself and her lab as bodily objects in the global spacetime context of the objective spacetime model of reality. In that case, it might happen that Wigner always discovers his measurement orientation corresponds exactly to his friend's orientation, so there was never any superposition involved and Wigner and his friend's outcomes always agree (yes, this is ridiculous).

As we noted, when it comes to Wigner's Friend some interpretations of quantum mechanics do entail the violation of intersubjective agreement (contradiction). For example, in Healey's pragmatic interpretation of quantum mechanics [7], it is the case that contradictions might obtain

[8]. In other words, the friend says she obtained +1 for the outcome of her qubit measurement while Wigner's outcome was (Friend)(–1). In yet other interpretations of quantum mechanics, the friend's information can be changed by Wigner's measurement.

For example, the Bohmian account of the more complicated Wigner's Friend experiment per Dustin Lazarovici and Mario Hubert [9] entails the following:

> the macroscopic quantum measurements of [Wigner] are so invasive that they can change the actual state of the respective laboratory, including the records and memories (brain states) of the experimentalists in it.

According to Lazarovici and Hubert, memories and records change, but the history of those memories and records (along their worldlines prior to measurement) remain intact, so nothing in the past is changed. It is analogous to passing vertically polarized light through a polarizer at 45° then sending it to a horizontal polarizer. The light incident on the first polarizer at 45° has no horizontal component, but it does after passing through the polarizer at 45°. Consequently, it can now pass through the horizontal polarizer. Thus, for the photons that are now passing through the horizontal polarizer, the polarizer at 45° can be said to have changed them from vertically polarized to (partially) horizontally polarized. Likewise, Wigner's measurements can literally change his friend's records and memories of her qubit measurement outcomes.

As previously noted in this chapter, earlier forms of relational quantum mechanics in its standard form also entail violations of intersubjective agreement in cases like Wigner's Friend. This is because the central idea in relational quantum mechanics is that different observers (they need not be conscious observers) may assign different, equally correct, quantum states to a given system. Again, this is because the quantum state assigned to a system describes not just the system itself, but also the relation between the system and the observer assigning the state. So, in relational quantum mechanics facts are relative to observers and the relation between any two systems A and B is independent of anything that happens outside these systems' perspectives, which means "two agents, both applying quantum mechanics correctly, [can] come up with incompatible predictions for the outcome of the same measurement" [1].

This can happen in the Wigner's Friend experiment because the friend says the wavefunction collapsed to a definite outcome in her "objective spacetime" (the spatiotemporal context of bodily/classical objects established via intersubjective agreement with all the observers in her lab) while Wigner keeps the friend and her lab in a superposition state until he has a measurement outcome in his "objective spacetime" (the spatiotemporal context of bodily/classical objects established via intersubjective agreement with all observers in the rest of the universe). The source of the contradiction is obvious, i.e., Wigner and his friend reside in different "objective spacetimes" [1]:

> [The friend] will perceive Wigner as saying that he has obtained the outcome [+1], even if Wigner himself thinks he has said that he has obtained the outcome [−1]. So not only will [the friend] and Wigner disagree about the outcome of a measurement, they will not even be able to rectify that disagreement by subsequently comparing notes. Wigner and [his friend] simply live within incommensurate realities now, and no attempt to reach across and bridge the gap can possibly succeed.

The existence of contradictory objective spacetimes (itself an oxymoron) is the result of extreme relationalism as epitomized by the use of the relative-state or "subjective collapse" formalism for quantum mechanics [3, 4]. In our ontology, there is only one objective spacetime so everyone must use the standard or "objective collapse" formalism for quantum mechanics [12].

Adlam and Rovelli recognized the problem and added a postulate called cross-perspective links to relational quantum mechanics [1]:

> In a scenario where some observer Alice measures a variable V of a system S, then provided that Alice does not undergo any interactions which destroy the information about V stored in Alice's physical variables, if Bob subsequently measures the physical variable representing Alice's information about the variable V, then Bob's measurement result will match Alice's measurement result.

However, this postulate creates an exception to the universal relationalism of relational quantum mechanics. We have shown how to keep the universal ontological contextuality advocated for by Rovelli without the kind of relationalism that leads to violations of intersubjectivity in the case of Wigner's Friend. That is, applying quantum mechanics to bodily objects and classical information, e.g., "Bob subsequently measures the physical variable representing Alice's information about the variable V," is where relational quantum mechanics differs from our view. One only needs 4D constraints ranging over quantum–classical contextuality to guarantee self-consistency and intersubjective agreement.

The bottom line, as shown above, is that none of these radical responses is required in the single, self-consistent objective spacetime model of reality presented here. What quantum mechanics and relativity are really telling us is that to exist as a bodily object in space and time is to interact with the rest of the universe directly or indirectly creating a consistent, shared set of classical information constituting the universe [11, 12]. Again, this is another way in which our view is a fully realist one, i.e., given that NPRF is at the core of our account of the physical world per quantum mechanics and special relativity [11, 12], The Absoluteness of Observed Events can never be violated in our view.

That is why in both Schrödinger's Cat and Wigner's Friend, the assumptions leading to the absurdity or contradiction are nonstarters in our view. Whenever one does measure a variable for some quantum, the outcome will become part of the self-consistent, shared information between observers satisfying intersubjective agreement on the model. The new scientific worldview given by our constraint-based, principle account of quantum mechanics guarantees it.

References

[1] E. Adlam and C. Rovelli, *Information is Physical: Cross-Perspective Links in Relational Quantum Mechanics*, 2022. Preprint. https://arxiv.org/abs/2203.13342.

[2] S. D. Bartlett, T. Rudolph, and R. W. Spekkens, *Reference frames, superselection rules, and quantum information*, Reviews of Modern Physics, 79 (2007), p. 555.

[3] V. Baumann and S. Wolf, *On Formalisms and Interpretations*, Quantum, 2 (2018), p. 99.

[4] J. Bub, *"Two Dogmas" Redux*, in Quantum, Probability, Logic: The Work and Influence of Itamar Pitowski, M. Hemmo and O. Shenker, eds., Springer Nature, London, 2020, pp. 199–215.

[5] D. Frauchiger and R. Renner, *Quantum theory cannot consistently describe the use of itself*, Nature Communications, 9 (2018), p. 3711.

[6] R. Healey, *Gauging What's Real: The Conceptual Foundations of Gauge Theories*, Oxford University Press, Oxford, 2007.

[7] R. Healey, *The Quantum Revolution in Philosophy*, Oxford University Press, Oxford, 2017.

[8] R. Healey, *IJQF Wigner's Friend Workshop*, 2018. https://www.ijqf.org/groups-2/workshop-on-wigners-friend-2018/forum/topic/is-there-an-inconsistent-friend/.

[9] D. Lazarovici and M. Hubert, *How Quantum Mechanics can consistently describe the use of itself*, 2018. Preprint. https://arxiv.org/abs/1809.08070.

[10] E. Schrödinger, *The present situation in quantum mechanics*, Naturwissenschaften, 23 (1935), pp. 807–812.

[11] M. Silberstein and W. M. Stuckey, *Re-thinking the World with Neutral Monism: Removing the Boundaries Between Mind, Matter, and Spacetime*, Entropy, 22 (2020), p. 551.

[12] M. Silberstein and W. M. Stuckey, *The Completeness of Quantum Mechanics and the Determinateness and Consistency of Intersubjective Experience: Wigner's Friend and Delayed Choice*, in Consciousness and Quantum Mechanics, S. Gao, ed., Oxford University Press, Oxford, 2022, pp. 198–259.

[13] W. M. Stuckey, *Schrödinger's Cat and the Qbit*, 2024. https://www.physics forums.com/insights/schrodingers-cat-and-the-qbit/.

[14] E. Wigner, *Remarks on the mind–body question*, in The Scientist Speculates, I. J. Good, ed., Heinemann, London, 1961, pp. 284–302.

Index